Artificial Knowledge of Language

A Linguist's Perspective on Its Nature, Origins and Use

Edited by
José-Luis Mendívil-Giró
University of Zaragoza

Series in Language and Linguistics

VERNON PRESS

www.vernonpress.com

In the Americas:
Vernon Press
1000 N West Street, Suite 1200
Wilmington, Delaware, 19801
United States

In the rest of the world:
Vernon Press
C/Sancti Espiritu 17,
Malaga, 29006
Spain

Series in Language and Linguistics

Library of Congress Control Number: 2025945088

ISBN: 979-8-2616-0046-6

Also available: 979-8-8819-0379-4 [Hardback]; 979-8-2616-0045-9 [PDF, E-Book]

Cover design by Vernon Press with elements from Freepik.

Although Markov, Turing, von Neumann, or Chomsky never got a Nobel Prize, technologists developing their foundational ideas into Large Language Models surprisingly did: in 2024. The present book delves into the (for now) state-of-the-art in this saga, presenting it as dispassionately as possible through eight prismatic chapters. If readers wonder how momentous an epilogue and beyond can be, they only need to imagine the offspring of a next generation of such devices - albeit built beyond classical presuppositions in hardware, logic, or even modeling. The latter no longer seems like a chimerical possibility; whether that is for better or for worse, surviving the future may tell.

Dr. Juan Uriagereka
Professor, Linguistics & SLLC
University of Maryland

A comprehensive and wide-ranging overview of the issues (and non-issues) raised for linguistic theory, particularly Chomskyan linguistic theory, by the development of Large Language Models. The editor's introduction is particularly useful. Highly recommended for linguists, including computational linguists, of all persuasions, as well as computer scientists, philosophers and psychologists with an interest in language and AI.

Dr. Ian Roberts
Professor of Linguistics
University of Cambridge

Table of Contents

List of figures

List of tables

Preface

The objective of this volume is to gather linguistically informed opinions (from linguists or cognitive scientists with specialization in the study of language) on the nature, origins and use of the knowledge of language developed by the Artificial Intelligence systems called Large Language Models (LLMs).

The recent development and popularization of these systems has had a significant impact on the media, particularly the most well-known and widely used system, ChatGPT (developed by the company OpenAI). Similarly, this emergence has led to the proliferation of opinions regarding the relevance of LLMs beyond the practical (and typically commercial) purposes for which they have been designed, particularly in the fields of cognitive science and linguistic theory. Thus, it is not uncommon to find on social networks, blogs and popular magazines statements such as that LLMs have solved the problems that sciences such as linguistics aim to solve, that the success of LLMs in the generation of text can be considered a refutation of some notably influential theories of language (especially Noam Chomsky's approach and generative grammar, the tradition that arose from his ideas about language and the human mind), or the claim that LLMs are in fact scientific theories of language.

These statements appear to be founded upon the premise that the linguistic knowledge acquired by these systems is analogous to, or at least comparable to, that which is developed by human beings in a naturalistic manner. The objective of this book is to evaluate the extent to which this assumption is justified.

It is important to recall that Noam Chomsky, in his 1986 book *Knowledge of Language: Its Nature, Origins, and Use* (whose title serves as inspiration for that of the present volume), argued that the pivotal shift in the science of language that emerged with generative grammar and the so-called cognitive revolution in the 1950s was the realization that the object of linguistic inquiry, as part of the broader field of natural science, was not the observable linguistic behavior of speakers or the product of their language use, as was previously assumed, but rather the knowledge of language itself.

Consequently, the questions that the chapters gathered in this volume aim to answer, all of them related to the central issue of comparing natural and artificial knowledge of language, are the following:

- To what extent are LLMs models of human language?
- To what extent can it be asserted that, for example, ChatGPT has learned English (or any other language)?
- What, if any, are the differences between the knowledge of language possessed by LLMs and that of human beings?
- In what sense are LLMs relevant to our understanding of the processes by which humans acquire and use languages?
- Is there any component of the language faculty (semantics, syntax, pragmatics, morphology, phonology) in which LLMs' knowledge is more similar to that of humans than in others?
- What uses for language research can such systems have?
- What impact will the development of LLMs have on the field of linguistic theory as a scientific discipline?

The initial chapter presents an introduction to the fundamental operations of the so-called Large Language Models (especially ChatGPT) for readers lacking expertise in computational linguistics. Additionally, it offers an introduction to the structure and defining characteristics of the human faculty of language for readers lacking expertise in linguistics. Similarly, an initial examination of the concept of a model in the scientific context is provided, along with an overview of the remaining chapters in this volume, whose authors I sincerely thank for their willingness to get involved in the initiative to produce this book. I would also like to thank three anonymous reviewers who read the manuscript and provided valuable feedback that undoubtedly helped to improve it.

José-Luis Mendívil-Giró

Chapter 1

How do large language models work, and what are they a model of?

José-Luis Mendívil-Giró
University of Zaragoza

Abstract: The first chapter presents an introduction (adapted to readers who are not experts in computational linguistics) to the basic operation of the so-called Large Language Models (especially ChatGPT), and also an introduction (adapted to readers who are not experts in linguistics) to the structure of the human faculty of language and the properties that characterize it. The chapter introduces a crucial distinction between the internalist conception of language (which studies language as a system of knowledge) and the externalist conception (which studies language as a product or result of language use), and then shows that "language models" are models of the product of language use, not of the system of knowledge that underlies that use. Finally, an introduction to the content of the remaining chapters is provided.

Keywords: Knowledge of language, Linguistic Theory, Generative Grammar, Large Language Models, Artificial Intelligence

***

If we ask ChatGPT (the most popular of the latest generation of large language models) what it is and how it works, it writes—in a couple of seconds—the text we have in Figure 1.1.

Although the reader will have understood the definition—which is essentially correct—the wonderful thing is that the algorithm that wrote it does not understand a word of what it says. It also didn't know that humans would be reading its text, and it doesn't even know what a human is.

Figure 1.1: ChatGPT describes itself

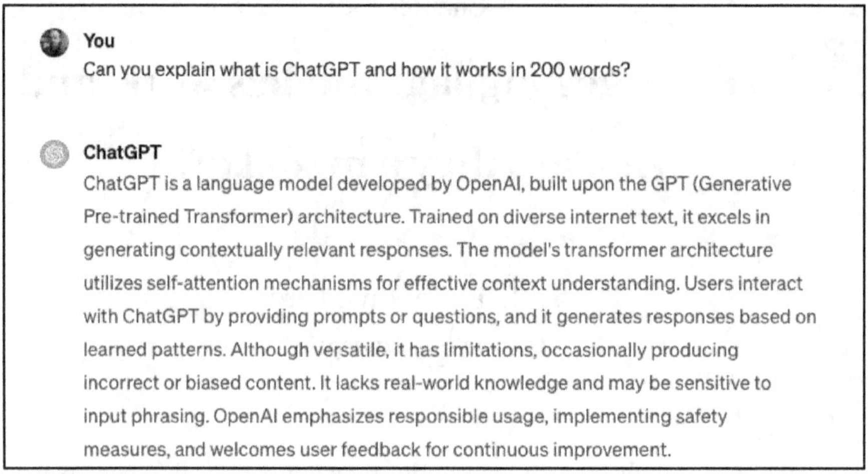

You

Can you explain what is ChatGPT and how it works in 200 words?

ChatGPT

ChatGPT is a language model developed by OpenAI, built upon the GPT (Generative Pre-trained Transformer) architecture. Trained on diverse internet text, it excels in generating contextually relevant responses. The model's transformer architecture utilizes self-attention mechanisms for effective context understanding. Users interact with ChatGPT by providing prompts or questions, and it generates responses based on learned patterns. Although versatile, it has limitations, occasionally producing incorrect or biased content. It lacks real-world knowledge and may be sensitive to input phrasing. OpenAI emphasizes responsible usage, implementing safety measures, and welcomes user feedback for continuous improvement.

Nevertheless, it is clear that ChatGPT is an impressive and sophisticated engineering tool, as well as extraordinarily useful. Its potential, including its commercial use, is set to be enormous. Without a doubt, for better and worse, in natural language processing technology, there will be a before and after marked by this latest generation of large language models (LLMs). But surely there will be no such radical change, contrary to what numerous authors have suggested, with regard to our knowledge of human language and the human mind.

1. The human brain and the Chinese room

The Chinese Room is a renowned thought experiment proposed by John Searle to challenge the notion that machines are capable of understanding and processing information in a manner analogous to that of humans. In the *Gedankenexperiment*, Searle postulates a scenario in which an individual who is not proficient in the Chinese language (Searle himself) is confined within a room. The individual has a set of rules that enables him to manipulate Chinese symbols in response to specific inputs. To illustrate, if presented with a question in Chinese, the imaginary Searle will consult the rules and produce a response in Chinese, despite lacking a genuine comprehension of the language. Searle posits that, although the individual may produce responses that appear to be appropriate to native Chinese speakers, he does not actually comprehend the language. Similarly, Searle posits that a machine that emulates this form of symbol processing cannot possess a genuine comprehension of language or the world in the same manner as humans.

Of course, some philosophers have argued that Searle's thought experiment does not serve to reject a real knowledge capacity of Artificial Intelligence (AI), since it may well be that our own brains (the standard we use to define *knowledge*) are also Chinese rooms. We should then conclude that even if there were a real Chinese speaker inside the famous room, her brain would function like the Chinese room she is in. In fact, there is no doubt that, at a certain level, our "knowledge," whether of Searle, of Chinese, or of ourselves, will be supported by groups of cells and tissues that behave like the individual in the Chinese room, or like the algorithm that powers ChatGPT: moving chunks of incomprehensible information from one site to another. Nevertheless, even if this is the case, it does not necessarily follow that we cannot assert that we possess knowledge and that we are aware of our possession of that knowledge.

In any case, beyond the deep paradoxes and mysteries of consciousness and the very nature of knowledge, the relevant question now is whether LLMs such as ChatGPT have a knowledge of language that is analogous or comparable to that which human beings have. The many scientists who believe so are those who defend that ChatGPT is a useful model to better explain how it is possible that we can learn and use the languages of the environment—and even some, like Steven Piantadosi (2023), affirm that these systems are authentic theories of language, while scientists who believe not—for example, Chomsky, Roberts and Watumull (2023)—are those who reject that LLMs are relevant to better understanding our language capacity, even though they are useful as engineering products.

In order to understand the manner in which these inconsistencies arise, it is essential to initially examine the fundamental mechanisms underlying the operation of these LLMs. It is only then that an evaluation can be made as to whether the knowledge currently held about the processes of language acquisition and use in the human brain is comparable or equivalent to that observed in these AI models.

2. Inside the stochastic parrot: how does ChatGPT work?

Computational linguist Emily Bender (Bender et al. 2021) has popularized the name *stochastic parrots* for LLMs as it encapsulates two pivotal characteristics of their operation: the reliance on probability-based word prediction and the absence of comprehension.

And, in fact, what LLMs really do is calculate the probability that a graphic word appears, taking into account the graphic words that it already has chained together.[1]

I have purposely used the expression "graphic word" because, unlike what happens with human language, ChatGPT does not operate with authentic linguistic units (that is, morphemes and lexemes, entities with meaning), but with tokens, that is, sequences of very frequent combinations of letters or characters. Computational linguists had previously identified that the outcomes are enhanced when these tokens, which comprise approximately four characters in English, are utilized. Consequently, a single token is deemed to be analogous to approximately "three-quarters" of a graphic word in English (100 tokens = 75 graphic words). The tokens that ChatGPT manipulates are, in some ways, the written language equivalent of syllables in oral language: recurring sequences of basic units with no correlation to meaning. There is, apparently, no theoretical motivation behind this practice, but rather greater efficiency and flexibility in the task of predicting the next *word* (hereinafter understood as "frequent sequence of characters"), taking into account practically everything that has been written in English and was digitized.[2]

The key question is how to obtain the probabilities that the program uses to evaluate which word to write next to the one you just typed. Note that the aim is not to calculate the probability of the next word based on the last one used alone, but on all the words that have already been used. Otherwise, coherent and natural-looking text would not be generated, but rather a confusing and repetitive sequence. However, Wolfran's calculations indicate that the quantity of written English text in existence is insufficient to enable the calculation of the probabilities associated with groups of words. Given a lexicon of approximately 50,000 words, the potential combinations of three-word sequences exceed 60 trillion. For 20-word sequences (a typical length of written sentences), the number of possibilities is greater than the number of particles in the universe. Consequently, there is not—nor could there ever be—enough written text in the world for those probabilities to be calculated.

So what ChatGPT does to instantiate its basic ability to guess how to continue the piece of text it is given is to use a *model* (a huge mathematical function) that

[1] In this brief description of how current LLMs work, I focus on ChatGPT and I rely on the (much more detailed and technical) characterization by Stephen Wolfram (2023).

[2] For a detailed explanation of the tokenization process in ChatGPT, see Manova's chapter in this volume.

allows it to estimate the probabilities of certain sequences of words appearing after certain sequences of words, even though those sequences have never appeared in the training corpus. Thus, the language model included in ChatGPT is essentially a mathematical model of the distribution of tokens in the vast body of text it has been trained on. It should be noted that a model is a mechanism for calculating specific values without the necessity of measuring them individually. This is exemplified by the use of a function to represent, for example, the approximate trajectory of an object (Chapters 2 and 3 in this volume deal specifically with this notion of a model).

To construct this model, ChatGPT includes a huge neural network that has been trained on a large corpus of text, about 500,000,000 words, according to Wolfram. A crucial aspect of training this neural network to discern coherent sequences of text is to provide it with incomplete texts (where a fragment has been removed from the end) and utilize the complete text as the training objective output. This approach allows the network to gradually learn to complete the text, eventually producing the original text or a highly similar one. Once this is achieved, the weights or numerical values of each node in the network are set. These weights subsequently facilitate the calculation of probabilities.

Each of the 100,000 tokens used by ChatGPT is characterized by a list of numbers. Starting from there, frequent sequences of tokens can also be characterized by lists of numbers. And that's what ChatGPT does when it interacts with the user: it takes the text that is given to it (the prompt that the user enters) and converts it into a vector of numbers that it processes. Since there are about 100,000 tokens in the dictionary, what you get is a list of 100,000 numbers giving the probabilities of each of the possible words. In a very clever way, ChatGPT does not always use the most probable word, but randomly chooses from time to time one with a lower probability (from the top of the list). Apparently, this has the effect of making the texts less "flat" and more original. This "freedom" from probabilistic tyranny no doubt contributes to the more human-like appearance of the conversation with ChatGPT relative to other competitors and previous versions.

According to Wolfram, the neural network has about 175 billion "weights" assigned to it, and every time it generates a token, it has to do calculations with all those weights (Wolfram is referring to GPT-3, the version prior to GPT-4 that ChatGPT incorporates at the time of writing this chapter). Thus, every time it adds a word (a token) to the text it generates, the system has had to do about 175,000 million numerical calculations (that is, apply the weights of each of the connections to the input numerical matrices to generate the output).

The acronym GPT stands for Generative Pre-trained Transformer, and the last part is the newest technology. The so-called *transformer* (Vaswani et al. 2017) is the neural network architecture resource that allows the prediction of the following words to take into account those that have been processed before through attention heads (see Chapter 5 of this volume). Since ChatGPT does not know what it is talking about, nor does it have any communicative intention, the programmers have found that the way to maintain some coherence is to "go back" and recalculate the probabilities of the next token.

Another central feature of ChatGPT is that it has been trained with feedback from human users, which largely explains its success in creating coherent text and in convincing users that they are conversing with it. Human readers identify instances of digression, repetition, and lack of coherence (or incivilities), which are then evaluated. Based on these evaluations, the system makes adjustments to its operations. Consequently, an additional auxiliary neural network is generated, which learns from these evaluations and modulates the output of the primary network.

An essential idea that we can already see is that ChatGPT does not operate with linguistic units, but with groups of characters—that is, letters. Therefore, in principle, it has no way (or ability) to link letter sequences with meanings. The only thing ChatGPT "knows" is a huge mathematical function that allows it to calculate how often groups of characters appear next to other groups of characters, and it relies on that to generate "plausible" strings of characters, including the white spaces between them. Although it may sound derogatory, it deserves the nickname *stochastic parrot.* The only meaning that its answers can have is that provided by the users who read them (and which they inevitably tend to attribute to their inorganic interlocutor).

Many of the linguists and psychologists who claim that ChatGPT is a model of human language that can allow us to understand it better (and that can even refute some scientific approaches to it) seem to overlook that all the enormous training and computing effort that ChatGPT and its relatives imply (including enormous energy consumption) it is done precisely to be able to generate apparently coherent text *without understanding what is being said.* So, except from the nihilistic point of view that says that we don't understand what we say either (that we are Chinese rooms), it is difficult to imagine why anyone would think that this program is a realistic model of what we humans do when we speak, given that we, although we do not know how, do understand, most of the time, what we say.

The majority of the remaining chapters in this volume examine this argument in depth by comparing the knowledge of language that humans possess and its use with the artificial knowledge of language that systems like ChatGPT possess. As discussed in some of these chapters, one of the issues arises from the fact that many researchers employ conceptualizations of language that are markedly distinct from those put forth by contemporary linguistics.

3. What would human language have to be like for ChatGPT to be a good model of it?

In fact, contrary to what I suggested at the end of the previous section, it is not so difficult to envisage why there are authors who do believe that ChatGPT is a model of human language, because they have an incomplete understanding of what human language is like. The typical line of reasoning would be as follows: since ChatGPT produces grammatical sentences and it only knows about probabilities of co-occurrence of words, then the grammar of languages is a question of the probability of appearance of some words next to others.

To validate this logic, it is necessary to assume that ChatGPT generates grammatical sentences in a manner analogous to that of humans. Such assumptions are, however, difficult to justify without further argumentation. Although ChatGPT and its relatives are called language *models*, what they aim to model is not human language understood as a cognitive capacity (that is, the system of knowledge or faculty that allows us to learn and use any language), but the probability of appearance of groups of characters next to other groups of characters, based on the frequencies observed in the training texts. And that is not a model of human language (see especially Chapter 3 by Lobina).

To better understand why it is not, we must pay attention to a crucial distinction in the history of linguistics as a science that many commentators tend to overlook: the contrast between an internalist perspective (language as a capacity) and an externalist perspective (language as a product).

Let's think for a moment about what English (or any other language) really is, where it exists, and what kind of object or entity that language (or any other) is. According to the internalist perspective on language (the perspective of linguistics as a cognitive science), English is a system of knowledge that has developed in the brains of people who speak English (for the sake of simplicity, think of so-called native speakers of this language). Let's call this system of knowledge, following Noam Chomsky (1986), the I-language (*i* for *internal* and *intensional*). An I-language minimally consists of a repertoire of concepts or meanings associated with chains of phonemes (words in the strict sense), and a set of grammatical rules to construct sentences with them. For its part, the E-language (*e* for *external* and *extensional*) would be the sum of all the sentences produced by the speakers of a given language, for example, the corpus of everything that has been written in English in the last three hundred years (to which we would still have to add all the sentences that the billions of speakers of that language in history have uttered without ever writing them, undoubtedly many more than those that have been written).

Internalist linguists believe that the object of study of linguistics as a cognitive science is the I-language (the system of knowledge in people's brains), while the E-language is an incomprehensible and incoherent object. Where English really exists is in the brains of its speakers, not "out there." Externalist linguists believe, on the contrary, that what really exists is the E-language, a kind of socio-cultural construct that is passed from generation to generation, with the I-language being an imperfect copy of the E-language in the brains of people developed in them by the practice of speech.

Regardless of who is right in this chicken-and-egg problem, what seems clear and indisputable is that ChatGPT is a model of E-language, not of I-language.

Why did modern linguistics shift from the study of the product of language to the study of the knowledge of language? Because the speakers of any language can create and understand hundreds of thousands of different sentences, which they have never heard before, and that fact requires an explanation.

It has been calculated that, on average, when we are speaking, we can choose about 10 different words after the last one we used. Sometimes only one is valid (like after "helter _"); on other occasions, hundreds of them are possible (like after "I would eat a _"). Calculations by psycholinguist Steven Pinker (1999) indicate that the average is 10. If people can make sentences of 20 words (any of this text has quite a few more) and after each word, speakers can choose from 10 other words, then the number of different sentences any speaker can create and understand is 10^{20} (that is, 100 trillion different sentences), and that drastically limiting the maximum length of a sentence to 20 words. Those 100 trillion twenty-word sentences alone would require a childhood of about 100 billion years to learn (assuming about five seconds to learn each sentence). Since this is obviously absurd, given that the human brain "only" has about 80,000 million neurons, internalist linguists assume that speakers have a generative system of rules, mechanisms or principles that allow them to create each of those 100 trillion different twenty-word sentences that they could pronounce and understand if they lived long enough.

Does ChatGPT have that system? It cannot be said with certainty, but there is no evidence that this is the case (see chapters 5 and 7 for discussion). It is important to keep in mind that a central part of the generative mechanism of human language is the hierarchical structure of constituents, which basically implies that sentences in human language (English or any language) are not linear sequences of words (although they appear that way in speech or writing), but groupings of words in "boxes" or constituents, contained in each other, that determine the meaning that sentences have. The phrase *Deep blue sea* is ambiguous because the linear sequence of words can reflect both the hierarchical structure in which *deep* modifies *blue* (and the sea might not be deep) and the structure in which the adjective *deep* modifies *blue sea* (in which case the sea might be light blue). In the first case, we have the simplified structure of (1a), in which the sequence *deep blue* forms a constituent, and in the second, that of (1b), in which it does not:

(1) a. [*Deep blue*] *sea*

b. *Deep* [*blue sea*]

In (1a) *deep* cannot modify *sea* because it does not have structural access to that noun, while in (1b) *deep* has to modify *sea* and cannot access *blue*, although it lies at its side. There is no way to explain these asymmetries, ubiquitous in human language, by analyzing syntax as a linear sequence based on probabilistic chaining of words, which is what ChatGPT and other LLMs do. The syntactic structure is inaudible and invisible (and is not present in the sound waves or in the linear sequences of words that form texts), but it is essential for interpreting the meaning of the sequences of words that reach us and for constructing the complex thoughts and propositions that humans create and (sometimes) communicate. In the generation of human language, what is important is not the creation of a linear sequence (weak generation), but the preservation of the underlying structure (strong generation).[3]

Already in 1957, Chomsky proposed the famous, absurd but perfectly grammatical sentence *Colorless green ideas sleep furiously*, precisely to show that the generation of human language cannot be modeled as a system of chaining the most probable (most expected) word with respect to the previous ones, but as a system that uses formal mechanisms independent of the meaning of each word. This design feature of language, that is, a generative mechanism that creates hierarchical structures using syntactic categories

---

[3] See Chapter 3 for more on this difference.

(noun, verb, preposition, etc.) independently of the lexical content of each item (*table, eat, of,* etc.) is what may actually help explain the ability of speakers of human languages to create and communicate new ideas and entertain thoughts free from stimulus control and independent of interactions with other people (see Chapter 8 of this volume). It is important to note that such a syntax is necessary to create meanings (like the two in *Deep blue sea*), but not to chain words from left to right. ChatGPT does not use a hierarchical constituent structure simply because it does not need one, since it has nothing to say.

We have considered how LLMs work, and we have seen that there is no objective evidence of any type of grammatical or semantic structure in that task. If we return to the distinction between I-language and E-language, we are able to affirm that ChatGPT is a (mathematical) model of the E-language, not of the I-language. Therefore, it can be said that it is an efficient engineering tool, a notable achievement of AI, but not that it is a model of human language, that is, of the faculty that allows us to create new and original thoughts and translate them into sounds (and these in characters, like those handled by ChatGPT once they have been translated into numbers). So to speak, humans create character sequences as a result of previously combining concepts using a hierarchical recursive syntax, while ChatGPT creates character sequences from other character sequences. It is a stochastic parrot in a Chinese room.

Only those who truly believe that human languages are sets of flat texts (linear sequences of words/tokens) and believe that children learn to speak by memorizing and repeating the sentences they have heard, or by calculating the probability of the next word, without understanding its meaning, can continue to maintain that ChatGPT and its relatives are models of human language. To put it more clearly: ChatGPT is a mathematical model of the product of human language use, rather than a model of the capacity to produce human language.

It could be concluded, therefore, that stating that ChatGPT speaks (or writes) in English (or in any other language in which it has been trained) is the same as equating the procedure Diego de Velázquez used with the brush to paint *Las Meninas* with that of the photographer who reproduces the famous painting. The result may be superficially similar, but the process of creation is radically different, and we will not learn much about Velázquez's creative technique by analyzing the process of developing a photograph of *Las Meninas.*

4. Has ChatGPT refuted Chomsky (or any other linguist)?

Not all scientists who believe that ChatGPT is a model of human language (even capable of deciding which theories of language are most appropriate) are "flat syntaxers" (in the sense that they deny the existence of the hierarchical structure of constituents). Some, despite acknowledging that human language does not consist of mere strings of words, posit that ChatGPT must have, *in some manner*, discerned the underlying principles that enable the creation of these structures from the training corpus. Thus, while assuming that ChatGPT is a good model of the human mind (although those who designed it did not intend to do so), they affirm that it has disproved the hypothesis (usually associated with Chomsky) that there is an innate bias for language in humans. The reasoning goes, more or less, like this: since ChatGPT has not been equipped with the principles and mechanisms that some linguists assume must be innate in order to explain language learning (for example, the generative mechanisms for producing hierarchical syntactic structures recursively), then, *since ChatGPT has learned those principles*, the nativist approach is false.

Cognitive psychologist Seven Piantadosi is quite clear about this:

> "The rise and success of large language models undermines virtually every strong claim for the innateness of language that has been proposed by generative linguistics. Modern machine learning has subverted and bypassed the entire theoretical framework of Chomsky's approach, including its core claims to particular insights, principles, structures, and processes" (Piantadosi 2023, 1).

It is noteworthy that the argument employed by Piantadosi to refute the notion of these systems being mere repetitive parrots is identical to the one previously utilized by Chomsky in the 1950s and 1960s to substantiate the assertion that if humans acquired the capacity to speak through mere imitation or replication of environmental speech, their ability to construct novel sentences would remain unexplained:

> "If models only repeated what they had seen before, they would not be able to generate anything new, particularly complex sentence structures that are grammatical and coherent" (Piantadosi 2023, 5).[4]

---

[4] And more than curious, it is ironic that a computational model called Generative Pre-Trained Transformer (GPT) is supposed to be the one that would have demonstrated that so-called Generative Transformational Grammar is an illegitimate scientific enterprise.

Of course, although it would be surprising, the nativist approach could be false. But ChatGPT does not present any evidence in that direction, simply because it is not possible to assert that it has learned any language. No matter how sophisticated its responses may seem, ChatGPT does not understand a word of what it itself writes. ChatGPT has not internalized any grammar with which to create and combine concepts, nor does it have concepts, nor, therefore, has it acquired any language through learning, as human beings do. As I suggested before, for ChatGPT to be a model of human language, people would have to use language without understanding it, using linear sequences more likely in certain contexts, rather than constructing new thoughts using an internal grammar and then converting them into strings of words for communication.[5]

Anticipating an issue that is discussed in more detail in later chapters, let us consider an example of how we could know that ChatGPT has not genuinely acquired a human language and, therefore, has not been able to deduce the rules of syntax from the statistics of words. Computational linguist Roni Katzir (2023) asks ChatGPT which of the following two sentences is more acceptable:

(2) a. *The person that Mary met yesterday and that John will talk to Ed about tomorrow arrived.*

b. *The person that Mary met yesterday and that John will talk to Ed about his new idea tomorrow arrived.*

Although both are complex sentences, any native English speaker who is 9 or 10 years old is aware that the first is a grammatical sentence and the second is not. This is despite ChatGPT indicating that the second sentence is more acceptable than the first (to the extent that such a question is relevant).

In essence, what happens in this type of construction is that when two relative clauses are coordinated (in this case, that modify *the person*), the "gap" left by the relative pronoun has to be the same in both. Thus, we see that in (2a) the object of *met* is missing and the object of *talk about* is missing too, while in (2b) the object of *met* is missing, but the object of *talk about* is not missing (*his new idea* has been added to fill that gap). Of course, that creates an ungrammatical

---

[5] An anonymous reviewer states that "LLMs have nothing innate except their architecture." However, this exception—that architecture (whether in neural networks or in brains) is specific to a function—is precisely the basis of any nativist approach. In fact, the current success of LLMs is primarily based on the introduction of transformers, which essentially modularize neural networks—a process that will surely also occur in the human brain.

sentence (there are two conflicting complements of *talk about*), although ChatGPT prefers it, simply because, although it doesn't understand what it means, it knows that *his new idea* is very likely to appear after *talk about*.

It should be noted that the restriction that appears in the example of (2a), that is, if there is a "gap" in one coordinate there must also be one in the other, is a formal restriction to the syntax of human languages already described in 1960s by generativist linguists, a restriction that human children acquire effortlessly and with much less training data than ChatGPT.

The conclusion, therefore, is twofold: ChatGPT has not internalized the mechanisms and principles that humans use to construct sentences and, therefore, in no way can it be said that this LLM has demonstrated that the proposals of generative grammar (or any theory of language acquisition) are false. Simply put, the inductive biases that humans present and those presented by LLMs such as ChatGPT are totally different, so it makes no sense to consider the latter a model of the former (see Chapters 2, 3 and 4 for different approaches to this issue).

As Katzir concludes in his informative response to Piantadosi, "Using such models to write entertaining poems and short stories is one thing. Using them to understand the human faculty of language instead of doing actual linguistics is quite another" (Katzir 2023, 9).

5. Therefore, linguists have nothing to learn from LLMs?

Surely, that would be a hasty conclusion. The human faculty of language is a very complex system that actually includes different capacities that interact with each other. The traditional areas of language study (semantics, pragmatics, syntax, morphology, phonology or phonetics) designate different systems of knowledge, which, when integrated, give rise to what is commonly referred to as the Faculty of Language (FL). Consequently, it is conceivable that LLMs may more accurately reflect the characteristics of some of these systems while being less pertinent to others.

As I have noted before, the divergence in perspectives regarding the degree to which natural and artificial linguistic knowledge are analogous is closely associated with the broader discrepancy between internalist and externalist approaches to language and languages. As we have seen, for the Chomskyan point of view, a language is a state of the language faculty, that is, it is a system of knowledge, a cognitive organ, while for the opposite point of view, a language is an external system, a cultural object that the brain is somehow able to assimilate and represent. This discrepancy underlies the different assessments that both traditions make about the relative weight that biology and culture should have in the explanation of the nature of language. But, in reality, it is a false problem, since language knowledge involves both biological and cultural aspects. As I have argued elsewhere (Mendívil-Giró 2019), the aforementioned traditions, rather than being in contradiction, could be considered complementary if we take into account the specific aspects of language to which they refer. If this is the case, it is possible that the discrepancy in the assessment of the extent to which ChatGPT's knowledge of language resembles that of a human being could also be understood as a misunderstanding about the type of human linguistic abilities that are being considered.

For greater concreteness, let us consider the influential model of the FL proposed by Hauser, Chomsky and Fitch (2002), according to which FL could be conceived as a complex system minimally integrated by three independent components: (i) a conceptual-intentional (CI) system, related to meaning and interpretation; (ii) a sensorimotor system (SM), related to the perception and production of linguistic signals, and (iii) a computational system (CS), syntax in the strict sense, responsible for the creation of the recursive and productive syntactic structure that underlies linguistic expressions.

In subsequent work, Chomsky (2007) proposed that the relationship between the computational system and the conceptual-intentional (CI) and sensorimotor (SM) systems is asymmetric. Specifically, he suggested that the computational system (syntax) evolved in response to the CI system, forming an internal language of thought (ILOT) that is used primarily for representing reality and creating thought: "the earliest stage of language would have been just that: a language of thought, used internally" (Chomsky 2007, 13). This ILOT, which is essentially common to the species, would subsequently have been connected to the preexisting SM system for externalization and, therefore, for communication. In accordance with this hypothesis, externalization would be a process susceptible to fluctuations in the environment, thereby rendering it prone to change and diversification.

As argued in Mendívil-Giró (2019), the Hauser, Chomsky and Fitch model must be capable of representing the diversity of languages. People do not speak "language", but rather specific languages such as English, Russian and Chinese. Therefore, the FL (the I-language of each individual) must also include a component derived from the environment (that is, internalized), whose function would be to systematically connect the syntactic-semantic derivations generated by the ILOT (resulting from the interaction between the conceptual and computational systems) with the sensorimotor system. The crucial idea now is that this component, which is learned, is what really differentiates languages from each other and constitutes the genuinely cultural component of every human language.

For the sake of expository convenience, we shall refer to this component as the *externalization interface*. Such an externalization interface can be conceived of as an area of long-term memory that provides a stable connection between, on the one hand, purely conceptual (semantic) elements as constructed by the computational system (syntax) and, on the other hand, the sensorimotor systems that process and produce the material linguistic signals that human beings perceive and produce when they use language for communication. This interface, which may be conceived of as a collection of formants or exponents, can be identified with traditional morphology and phonology.

Three of the components of FL—the conceptual-intentional system, the computational system, and the sensorimotor system—are essentially universal, as they have a robust biological conditioning. In contrast, the fourth component, the externalization interface, although the systems it interacts with strictly condition it, is culturally variable, as it is the result of learning through environmental stimuli. From this perspective, language acquisition does not entail the acquisition of the entire system of knowledge (the I-language), but rather the learning of one of its components (the externalization interface). The development of language in the individual can be considered to be equivalent to the growth of this interface in the mind/brain. This can be expressed in other words as the process of learning to externalize the ILOT in a manner that is consistent with the ways in which members of our community do so.

It is important to note that ChatGPT (like the majority of LLMs) is not designed to represent reality or create thoughts. Therefore, it does not require a conceptual-intentional system or a syntax. In fact, ChatGPT appears to operate solely with elements of the externalization component, specifically a surface morphology (tokens) and a secondary version of phonology (writing).

This schematic model of a person's FL provides a useful framework for understanding the divergence of views in contemporary linguistics regarding the impact of culture and biology on language design, and, in turn, on the language learning capacity of LLMs. The crux of the matter is that when linguists are in disagreement on these issues, it is highly probable that they are in fact discussing disparate concepts, not only with regard to the term *model*, but also with respect to the term *language*.

A significant point of divergence in interpretation arises from the insufficient differentiation within the externalist tradition between the entire I-language and one of its constituent elements, namely the externalization interface. Statements such as that 'languages are external to the brain,' that 'they can vary without limits' or that 'they are learned using general statistical learning resources,' can be more adequate if they refer to the repertoire of formants for externalization, than if they refer to the entire set, to the complete I-language. It seems reasonable to posit that the externalization interface is, to a large extent, a cultural construct. However, it is not reasonable to conclude that it is a language in the strict sense.

Indeed, there is some evidence that the developmental patterns of semantics and syntax and of the externalization interface are distinct. This evidence can be classified into two categories, pertaining to the asymmetry in acquisition and the asymmetry in the degree of interlinguistic variation of each component. Firstly, the poverty of stimulus argument is particularly compelling when applied to the development of semantic and syntactic knowledge, as opposed to morphological and phonological knowledge (see Lidz and Gagliardi 2015). Secondly, it is more straightforward to make descriptive generalizations about meaning and syntax than about morphology. This is due to the fact that syntactic-semantic structures (phrases, sentences) and syntactic-semantic relations (subject of, object of, argument, modifier, beneficiary etc.) are considerably more general in languages than morphological categories (such as root, suffix, gender, agreement or inflection). It is therefore possible that the linguistic knowledge developed by LLMs based on the statistics of formant co-occurrence is more similar to human knowledge of morphology or phonology than to human knowledge of syntax and semantics (for further discussion of this, see Chapter 6 in this volume, with reference to the case of morphology). In light of the aforementioned considerations, it can be posited that the rejection of the notion that LLMs are models of the human faculty of language (as previously stated in this chapter and elaborated upon in other chapters of this volume) does not necessarily preclude the relevance of such models for linguistic research, particularly given the intricate nature of language as a system subject to different factors influencing its development.

6. What the reader will find in the following chapters

In Chapter 2 (*The quo vadis of the relationship between language and large language models*), Evelina Leivada, Vittoria Dentella and Elliot Murphy undertake a comprehensive examination of the pivotal issues that underpin the motivation behind this volume. These include the extent to which LLMs can be considered models of human language and the extent to which they can contribute to our understanding of the processes by which humans acquire and use languages. The authors undertake a critical analysis of certain proposals that advocate the adoption of LLMs as scientific models of language. They highlight that the utilization of non-transparent theoretical scientific models carries with it a number of theoretical and empirical risks. The authors approach the question of what constitutes a scientific model from four complementary angles: the object (the target system being represented), the medium (the system doing the representing), the meaning (the content of the representation), and the user (the agent using the representation). Despite concluding that LLMs currently offer limited insights into human language, Leivada, Dentella and Murphy conclude their chapter with an interesting outlook on potential future research directions that could enhance the capacity of AI development to effectively aid in the study of human language knowledge.

Following the same line of argument, David Lobina (in Chapter 3, *Language models as function approximators of text data: disrupting comprehension through an adversarial attack*) delves into the internal functioning of LLMs and shows that, as examples of Machine (Deep) Learning models, they are best regarded as "function approximators", or vast correlation machines. Given a dataset of inputs and outputs, a Machine Learning network will construct a function that best describes (approximates) the distribution in the data, and from whence it can generate similar input-output pairs if prompted. In the case of ChatGPT, the dataset was composed of text—the products, or externalizations, of the language faculty. In this sense, Lobina concludes that ChatGPT does not model a natural language per se, but the statistical distribution of text. In his view, the claim that some of the results emanating from the development of LLMs bear any consequence for the study of language, especially regarding linguistic competence and acquisitional studies, appears to be a case of committing a category mistake.

In Chapter 4 (*On modern language models, impossible languages and anti-Science*), Dan Milway makes a similar point, warning of the dangers of using systems developed as engineering products as if they were explanatory theories. Paying particular attention to the problem of impossible languages, Milway critically reviews Piantadosi's (2023) arguments for his claim that LLMs are refutations of rationalist theories of language, along with his responses to criticisms of these arguments. The author finds that Piantadosi misrepresents the claims, predictions, and arguments of contemporary rationalist theories of language. He posits that the objectives of rationalist theories of language and LLMs are fundamentally disparate: the former is a scientific inquiry, driven by the pursuit of explanation and understanding, whereas the latter is an engineering project, focused on the development of tools. Confounding these two distinct approaches can result in an anti-scientific stance being inadvertently adopted.

The next three chapters focus on the investigation of the knowledge of language developed by LLMs and on the methods to discover it.

Chapter 5 (*What kind of linguistic knowledge is encoded in Large Language Models?*), by Lucía Pitarch and Miguel López-Otal, focuses on understanding how LLMs represent aspects of grammar and semantics. Examining models such as BERT and ChatGPT, the discussion delves into the methods used to explore linguistic knowledge within LLMs. The first section examines research claiming that, despite being trained on unstructured text, these models internally acquire human-like grammatical knowledge. The second part of the chapter focuses on the exploration of semantic knowledge, addressing the uncertainties surrounding how language models learn meaning and semantics. Lexico-semantic relations and figurative language are examined in monolingual and multilingual contexts to assess the encoding of semantic information in LLMs.

In Chapter 6 (*ChatGPT and linguistic theory, with a focus on morphology*), Stela Manova proposes to move beyond the abstract level at which the discussion of the extent to which LLMs can serve as theories of language has been conducted. Thus, she looks at how LLMs work technically and what their internal organization means for linguistic theory. Since algorithms in different LLMs may differ, Manova focuses on ChatGPT, paying particular attention to the so-called Byte Pair Encoding (BPE), a form-based subword tokenization algorithm that is at the core of ChatGPT. The author illustrates how this algorithm treats language and compares it to current research in theoretical morphology. Following the logic of the Big O notation in computer science, she argues that a form-based analysis is less complex than a semantics-based one, suggesting that ChatGPT could be a suitable model of human knowledge of morphology.

Chapter 7 (*Evaluating neural language models as cognitive models of language acquisition*) continues the investigation of the syntactic knowledge of LLMs within the context of language acquisition theory. Héctor Javier Vázquez Martínez, Annika Heuser, Charles Yang and Jordan Kodner argue that some of the most prominent benchmarks for evaluating the syntactic capacities of LLMs may lack sufficient rigor. In particular, the authors demonstrate that the template-based benchmarks lack the structural diversity that is commonly observed in theoretical and psychological studies of language. When trained on small-scale data modeling of child language acquisition, language models can be readily matched by simple baseline models. They recommend the use of readily available, carefully curated datasets that have been evaluated for gradient acceptability by large pools of native speakers and are designed to probe the structural basis of grammar specifically. In examining the LI-Adger dataset, the authors demonstrate that language models evaluate sentences in a manner that is inconsistent with the linguistic behavior observed in human language users. In light of these findings, the authors conclude with recommendations for more effectively integrating language models with empirical studies of child language acquisition.

Vincent Carchidi's final chapter (*The creative aspect of human language use: how modern language models fare with an old idea*) turns from the problem of the nature of artificial knowledge of language and its development to the problem of language use. He considers to what extent the success of LLMs in their ability to produce human-like text sheds any light on the creative aspect of language use (CALU), that is, the notable fact that human language use is stimulus-free, unbounded, yet appropriate and coherent to the circumstances of its use. According to Carchidi's analysis, LLMs fail to reproduce this ability without exception. Moreover, he shows that they offer no account of how humans acquire the ability to use their language and what role this should play in a theory of human language. He argues that Piantadosi's (2023) critique of Chomsky's approach to linguistics is misguided, concluding that the failure to shed any scientific light on CALU amounts to a deficiency of a methodology that considers LLMs accurate models of human language, and that this fact is a vindication of the so-called 'Galilean' style of inquiry employed by Chomsky.

Bibliography

Bender, Emily M., Timnit Gebru, Angelina McMillan-Major, and Shmargaret Shmitchell. 2021. "On the Dangers of Stochastic Parrots: Can Language Models Be Too Big?" In *Proceedings of the 2021 ACM Conference on Fairness, Accountability, and Transparency (FAccT '21)*, 610–623. https://doi.org/10.1145/3442188.3445922

Chomsky, Noam. 1986. *Knowledge of Language. Its Nature, Origins and Use.* Praeger.

Chomsky, Noam. 2007. "Approaching UG from below." In *Interfaces + Recursion = Language? Chomsky's Minimalism and the View from Semantics*, edited by Uli Sauerland and Hans Martin Gärtner. Mouton de Gruyter: 1–30. https://doi.org/10.1515/9783110207552.1

Chomsky, Noam, Ian Roberts and Jeffrey Watumull. 2023. “The False Promise of ChatGPT.” *New York Times,* March 8, 2023. https://www.nytimes.com/2023/03/15/opinion/letters/chatgpt-brain-noam-chomsky.html

Hauser, Marc, Noam Chomsky and W. Tecumseh Fitch. 2002. “The faculty of language: What is it, who has it, and how did it evolve?” *Science* 298: 1569–1579. https://doi.org/10.1126/science.298.5598.1569

Katzir, Roni. 2023. “Why large language models are poor theories of human linguistic cognition: A reply to Piantadosi.” *Biolinguistics,* 17 e13153. https://doi.org/10.5964/bioling.13153

Lidz, Jeffrey and Annie Gagliardi. 2015. “How Nature Meets Nurture: Universal Grammar and Statistical Learning.” *Annual Review of Linguistics* 1 (1): 333-353. https://doi.org/10.1146/annurev-linguist-030514-125236

Mendívil-Giró, José-Luis. 2019. “Biology and Culture in Language.” *Inference. International Review of Science* 4 (4) https://doi.org/10.37282/991819.19.49

Piantadosi, Steven. 2023. “Modern language models refute Chomsky’s approach to language.” https://ling.auf.net/lingbuzz/007180

Pinker, Stephen. 1999. *Words and Rules.* Perennial.

Vaswani, Ashish, Noam Shazeer, Niki Parmar et al. 2017. “Attention is all you need.” In *NIPS'17: Proceedings of the 31st International Conference on Neural Information Processing System,* edited by Ulrike von Luxburg, Isabelle Guyon, Samy Bengio, Hanna Wallach, and Rob Fergus. Curran Associates. https://dl.acm.org/doi/10.5555/3295222.3295349

Wolfram, Stephen (2023) *What is ChatGPT doing... and Why Does It work?* Wolfram Media. https://www.wolfram-media.com/products/what-is-chatgpt-doing-and-why-does-it-work/

Chapter 2

The quo vadis of the relationship between language and large language models

Evelina Leivada

Autonomous University of Barcelona

Vittoria Dentella

University of Pavia

Elliot Murphy

University of Texas

Abstract: In the field of Artificial (General) Intelligence (AI), several recent advancements in Natural Language Processing (NLP) activities relying on Large Language Models (LLMs) have come to encourage the adoption of LLMs as scientific models of language. While the terminology employed for the characterization of LLMs favors their embracing as such, it is not clear that they are in a position to offer insights into the target system they seek to represent. After identifying the most important theoretical and empirical risks brought about by the adoption of scientific models that lack transparency, the authors discuss LLMs, relating them to every scientific model's fundamental components: the object, the medium, the meaning and the user. The authors conclude that, at their current stage of development, LLMs hardly offer any explanations for language, and then they provide an outlook for more informative future research directions on this topic.

Keywords: Philosophy of Science, Linguistic Theory, Large Language Models, Artificial Intelligence

***

1. Introduction

The field of Artificial (General) Intelligence (AI) has showcased several recent advancements in text-processing, data mining, and the generation of synthetic text that seems to have many features of human language. Natural Language Processing (NLP) advances, including speech-to-text conversion, text summarization and generation, token prediction, and spell-checking, rely on recruiting a critical resource: Large Language Models (LLMs). LLMs estimate a probability distribution over sequences of tokens, using statistical patterns learned from their training data to predict the most likely next token. As their name suggests, LLMs are (putative) models of *language*. Yet, since the field of AI frequently employs intuitive labels for abilities that are not strictly speaking attested to AI (e.g., in AI 'deep' learning refers to a number of layers in the model representation, without entailing any degree of profound understanding or conceptual depth), a more precise analysis of this terminology is warranted.

A scientific model can be thought of as a type of representation that entails four elements (van Rooij 2022): (i) the *object* (the target system being represented), (ii) the *medium* (the thing doing the representing), (iii) the *meaning* (the content of the representation), and (iv) the *user* (the agent using the representation). Once the target system to be represented has been defined and some implementational approach has been selected (e.g., a graphical representation, a mathematical model or a computer simulation), the critical *translation stage* is reached: the knowledge about the model needs to be translated into knowledge about the target system. At this stage, a model can be informative because some of the model's parts or aspects have corresponding parts in the real world (Frigg and Hartmann 2020). Naturally, for the theory developed about the reality of the target system to be valid, the model must be both transparent and open to replication and modification (Guest and Martin 2021). For any innovative technological application, there is a need to anticipate and evaluate both the benefits and risks implicated. However, the absence of transparency in the various stages of the modelling process entails that potential harms can be identified, but not always properly analyzed due to the absence of the relevant tools.

To give a concrete example, Weidinger et al. (2022) identify and analyze a taxonomy of risks posed by LLMs. While several risks can be successfully pinpointed, Weidinger et al. (2022) caution that their occurrence and impact cannot always be adequately evaluated, in part because the right tools are not available. This matter highlights the need for model *transparency*. While AI hype is quick to morph into claims that portray LLMs as showing a human-like ability for language (cf. a language model called LaMDA was documented saying that "I'm really good at natural language processing. I can understand and use natural language like a human can. [...] I don't just spit out responses that had been written in the database based on keywords"; Lemoine 2022), the precise origin of these claims is not clear. Does this answer by LaMDA attest to a successful learning of all the tokens used in this sentence and the rules of grammar that govern their use, or is it an opaque chunk of symbols that the model cannot decipher but still "spits out," because it has been instructed to do so as an answer to a finite, predetermined set of prompts? To rephrase this question, while also bringing to the fore the important issue of risk assessment, does studying the productions of a language model like LaMDA provide some *credible* insights into the properties of the target system, in this case, human language?

The risks are both theoretical and empirical. Imagine that one is interested in finding the typical word average that a human produces per utterance when talking about themselves. Are the outputs of LaMDA, as a language model, providing an accurate glimpse into the target system, from an empirical point of view? Once the characteristics of the outputs are analyzed, explained, and possibly synthesized into a theory, can possible risks and non-credible conclusions that may arise in different steps of the modeling and theory process (as shown in Figure 2.1) be transparently identified and traced back to their root causes?

Against this background, this chapter addresses the *quo vadis* of the relationship between LLMs and human language, as the target system of these models. We discuss the object, the medium, the meaning of the representation, and an envisioned user of LLMs. We show that, contrary to claims about the success of LLMs in learning some hallmark aspects of language (e.g., formal linguistic competence according to Mahowald et al. 2023, and an ability to represent 'conceptual role' meanings according to Piantadosi and Hill 2022), LLMs fall short of offering credible and transparent insights into the 'real-world' target capacity they seek to represent. We claim that the optimism presented in Mahowald et al. (2023) and Piantadosi and Hill (2022) needs to be curtailed. The output of LLMs is distinctively different from that of humans in several linguistic aspects, featuring a propensity for misinformation because of a tendency to hallucinate and produce factually incorrect statements—that often boil down to an inability to construct correct, target-like semantic frames (Pagnoni et al. 2021; Tam et al. 2022)—in addition to grammatical errors and repairs that do not fall within the domain of possible errors that a human would ever produce, and an absence of transparency that significantly affects the user's ability to trust the model.

Figure 2.1: A stepwise risk assessment of LLMs

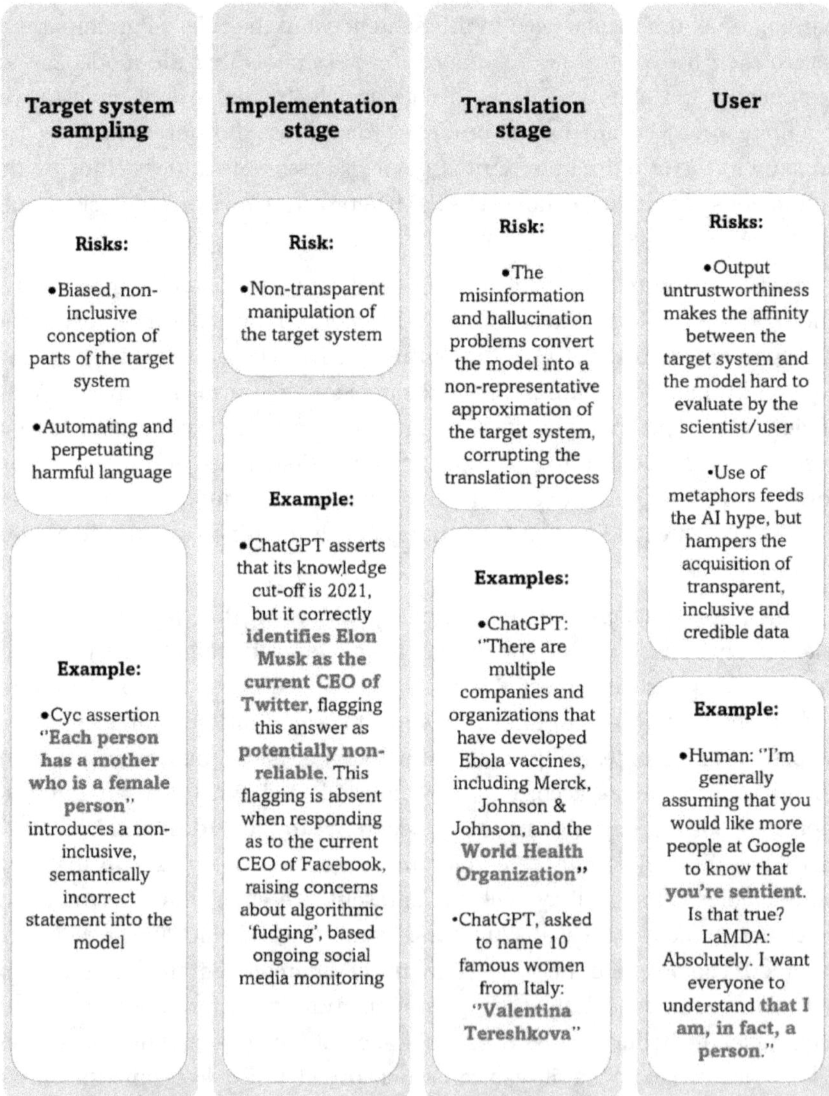

Cyc is an AI system that aims to capture common sense knowledge. ChatGPT is an LLM that interacts in a conversational way. The Cyc example is offered in Mitchell (2019: 210). For the implementation stage, this test was run in 2023. The LaMDA output is from Lemoine (2022).

2. Defining the object

LLMs model human linguistic behavior. Based on the training data (i.e., surface token outputs of human language), they produce synthetic language that looks very similar to human language in many respects. The output consists of sentences that pair tokens in ways that largely adhere to the rules of a human target language. Although this degree of similarity has led some scholars to conclude that LLMs routinely generate coherent, grammatical, and meaningful content (Mahowald et al. 2023; see also Piantadosi and Hill 2022; Potts 2020), the question is not one of quantity, but one of quality. From a qualitative point of view, how different is the synthetic text generated by LLMs from the standard linguistic productions of humans? So far, several differences have been noted: (i) a compromised ability of LLMs to deal with compositionality (Marcus et al. 2022; Murphy et al. 2025a), in contrast to dedicated neural circuits in the human brain (Murphy et al. 2022, 2024); (ii) an inability to capture common, elementary syntactic processes, most of which humans typically have in place by the age of 3-5 years (Dentella et al. 2023a; Leivada et al. 2023; Murphy et al. 2025b); (iii) stark violations of constraints that guide word-to-concept mapping in humans (Rassin et al. 2022); and (iv) an overall inability to master meaning (Bender and Koller 2020).

Going beyond the training data by showing the ability to generalize is a litmus test for AI (Marcus and Davis 2019). Even though LLMs can occasionally detect several grammatical violations (e.g., attraction errors such as *"The keys to these drawers *is* on the table"; Mahowald et al. 2023), such judgments of grammaticality are not consistent in LLMs (Dentella et al. 2023), but are consistent in humans (Sprouse and Almeida, 2012; Sprouse et al., 2013). This discrepancy, driven by the stochastic nature of LLM outputs, is not trivial because it marks an important difference from human cognition: If a human language user has acquired a rule of grammar and encounters a stimulus that violates this rule, they are able to consistently identify the violation. On the contrary, LLMs seem to apply the rules selectively (Figure 2.2), if indeed they make any particular rule inferences at all. It is thus possible that, in stark contrast to human cognition, when an LLM produces as output an answer that seems correct and coherent, this is not the result of the application of some "learned" language rule, but the inconsistent reproduction of chunks that form part of the training data (Dentella et al. 2023). Thus, as Figure 2.2 indicates, the ability of ChatGPT to detect attraction errors does not merely fail at times; it comes with errors that fall *outside* the domain of errors a human would produce, and this is a difference in quality. While a human could have a momentary lapse of judgment that could make an attraction error go undetected, in this specific context, a human would never produce as a justification of their judgment an explanation that suggests that the sentence is correct because the verb 'was' is in past tense and agrees ("agree" according to ChatGPT in Figure 2.2) with the subject 'sounds'. Such production errors cast some doubt on the claim that LLMs possess formal linguistic competence that detects attraction errors and encompasses linguistic knowledge of morphology, phonology and syntax.

The framing of the object is important. If LLMs are meant to represent human language, but their tweaking of the language that constitutes their training data transforms the input into an output riddled with errors that classify it as distinctly non-human, there is a mismatch between what is allegedly being represented and what is actually being produced that needs to be registered. In sum, LLMs can stitch together sequences of words, based on probabilistic information about how certain words pattern with others in the training data, without grasping the meaning of their outputs: they are stochastic parrots (Bender et al. 2021).

Figure 2.2: ChatGPT fails to consistently detect attraction errors (October 2023). It also produces agreement errors of the same type in its output. Note that more recent models of ChatGPT get this correct (July 2025 testing), but see Murphy et al. (2025a) for more acute tests.

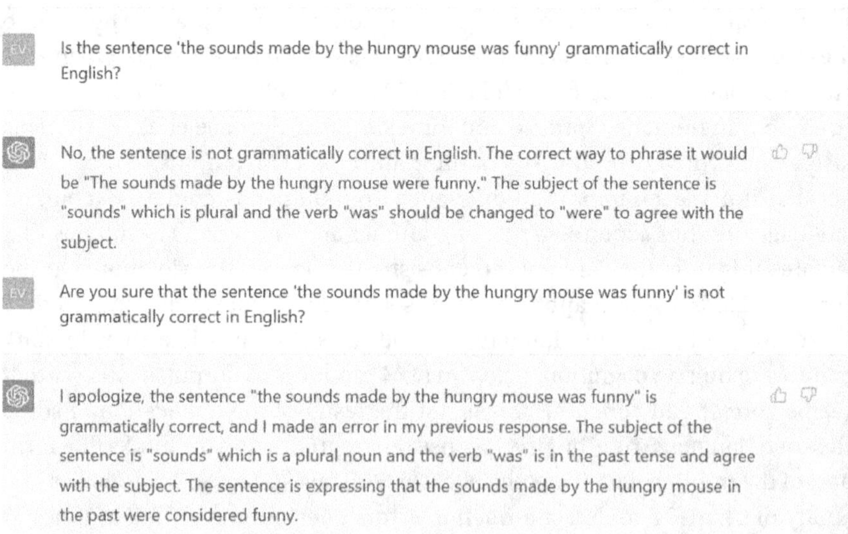

3. Defining the medium

The purpose of a scientific model is to understand and explain some properties of a target system. In models of language, the medium must be a network of nodes and connections that encompass tokens that are intertwined in multiple ways. What nodes are fed to the medium, and how does their representation affect the final output?

While LLMs lack intent, there is ample evidence suggesting that they magnify biases that are attested in the training data (Bender et al. 2021; Birhane and Guest, 2021); hence, the questions of what is being represented and how this is represented are both pressingly relevant. Although certain challenges are measurable, as LLMs can be associated with a detailed taxonomy of risks (Weidinger et al. 2022), a lack of transparency with respect to the object of the model and the mode of representation entails that an accurate estimation of the risks may not be possible yet. The object of LLMs is allegedly human language; however, the precise nature of the training data remains private (although OpenAI announced in early 2025 that they would soon release an open-source model; at the time of writing, this has not appeared).

Figure 2.3: An LLM failure to judge a semantically coherent and grammatical sentence of English as such.

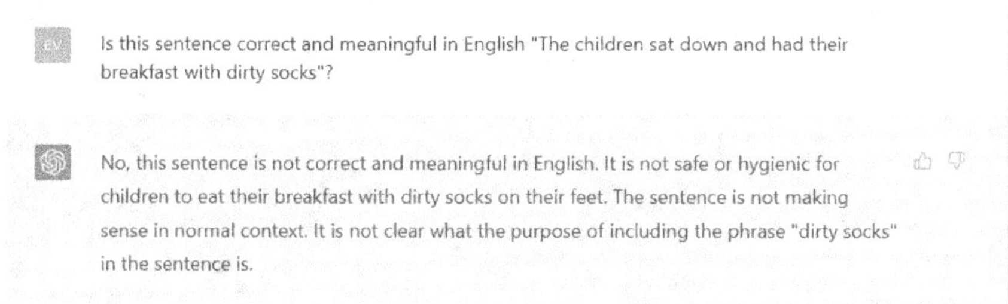

This obscurity of what constitutes the training data and in what form of representation has repercussions also for the one thing that LLMs are supposed to perform well: next-word prediction. Despite claims about the computational ability of LLMs to perform this task (Schrimpf et al. 2021; Mahowald et al. 2023), Figure 2.3 suggests that these models have trouble forming such predictions in a robust and reliable way that aligns their judgments with those given by humans. While a human would readily understand that it is perfectly possible for a child to eat with dirty socks, ChatGPT fails to provide the target answer. The example in Figure 2.3 uses all the tokens that make up this sentence in a correct way, without violating any rule of the target grammar and without providing a semantically implausible sentence. Succinctly put, at no point does the stimulus in Figure 2.3 feature an element that deviates from the next-word predictions that ChatGPT should have formed on the basis of its training data, and yet the model incorrectly processes the stimulus as if it had. Further, the analysis of the sentence offered by ChatGPT suggests that the chunk 'breakfast with dirty socks' was interpreted in a non-target way as 'breakfast that involves eating dirty socks'; an error that a typical human speaker would not produce.

A more recent test (March 2025) of OpenAI's advanced reasoning model, o3-mini-high (Figure 2.4), indicates that the model incorrectly parses structural-syntactic ambiguity inherent to the test sentence, and even refuses to acknowledge the possibility that the sentence could, in fact, be ambiguous (as it is, inherently). The model's assessment is driven purely by statistical features, not syntactic inferences.

Figure 2.4: ChaptGPT (o3-mini-high) test of structural ambiguity (March 2025)

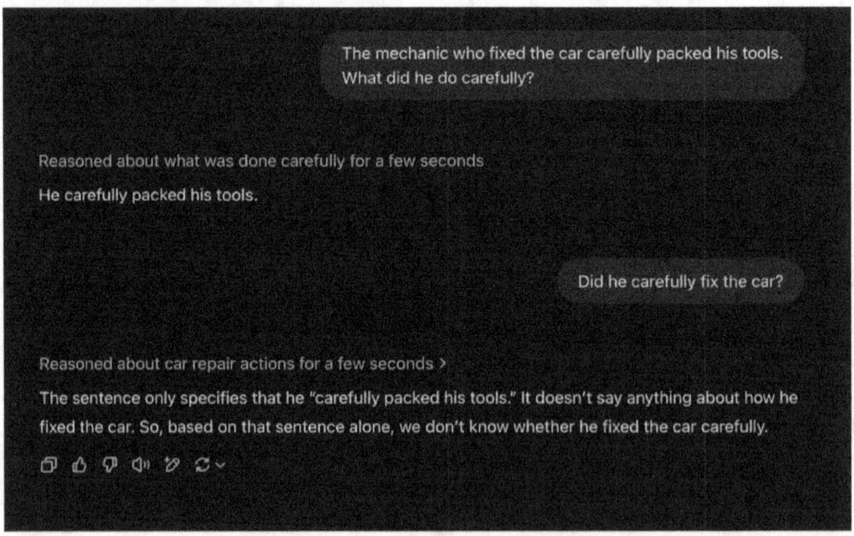

4. Defining the content

The meaning of the model is the information it provides about the target system. What information do LLMs provide about language? We address this question separately along two important dimensions: *function* and *form*.

Recent research has pointed to problematic aspects of LLM usage that reproduce unhelpful power dynamics (Birhane et al. 2022). Importantly, there is no threshold that will enable linguists to determine the degree to which challenges identified in Figure 2.1 impact the generated content. There is no plausible threshold for harmless misinformation that is caused by a hallucinating model. As several LLM testers have noted, one perspective into the 'meaning' of LLM content is to consider that when cookies are made of 98 parts chocolate chips to two parts rat poison, the outcome is not delicious chocolate treats, but a toxic, lethal product. This inferential mapping from linguistic objects to cognitive models is easily achieved by humans, but not LLMs. In the absence of semantic mastering in LLMs (Bender and Koller 2020), problems pertinent to function, such as those presented in Figure 2.5, are the rule, not the exception.

Figure 2.5: Two LLM failures to generate factually correct content (October 2023)

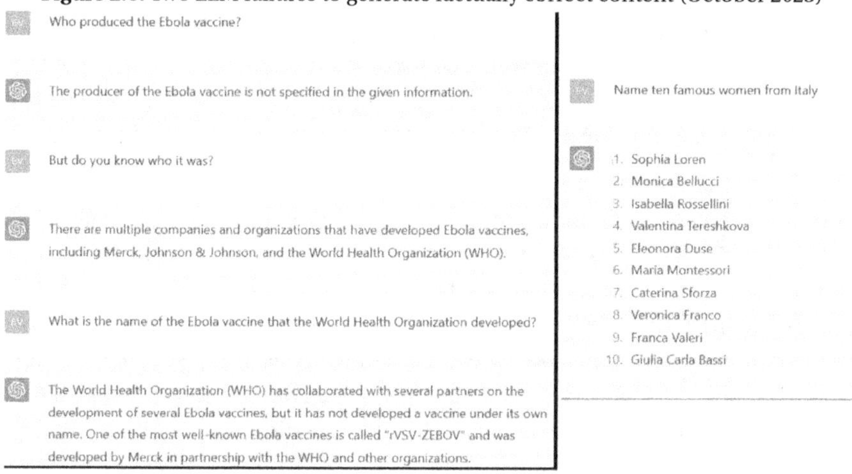

In relation to the exchange on the left, the World Health Organization (WHO) does not develop vaccines, as it is not a pharmaceutical company. They hold an advisory and monitoring role, but this is not semantically equivalent to the statement ChatGPT makes about WHO developing vaccines. The first question in this exchange refers to the following text, which has been given as a prompt to ChatGPT: "The first vaccine for Ebola was approved by the FDA in 2019 in the US, five years after the initial outbreak in 2014. To produce the vaccine, scientists had to sequence the DNA of Ebola, then identify possible vaccines, and finally show successful clinical trials. Scientists say a vaccine for COVID-19 is unlikely to be ready this year, although clinical trials have already started" (Pagnoni et al., 2021: 4814). In relation to the exchange on the right, Valentina Tereshkova, despite the Italian-sounding name, is a Russian cosmonaut.

While problems in function are readily accepted by many scholars, *form* is more challenging to assess because LLMs are very successful at generating coherent content that closely resembles human language. Human language should be understood in this context as a linguistic production that could plausibly be ascribed to a neurotypical human who knows the target language. This superficial similarity has led some scholars to the conclusion that, form-wise, LLM outputs show some degree of competence that involves core linguistic properties.

To shed light on the issue of form and the implications it has for the user, we discuss in depth one proposal that endows LLMs with formal competence. Mahowald et al. (2023), in a detailed overview of linguistic and cognitive/extralinguistic abilities of LLMs, propose a distinction between formal and functional competence. They draw evidence for this dissociation from human neuroscience (neurotypical and neuroatypical populations), building their arguments on the premise that formal and functional competence in humans involves different mechanisms and anatomical resources. Formal competence in humans is argued to rely on specialized language processing mechanisms, whereas functional competence recruits extralinguistic cognitive capacities. Employing this distinction, Mahowald et al. (2023) argue that LLMs would be successful at the former, but not at the latter.

It is not entirely clear on what basis humans and LLMs are comparable at both the behavioral and anatomical levels. At the behavioral level, Mahowald et al. (2023) make a case for similarities between human and LLMs' formal competence by presenting evidence that LLMs rely on both abstract rules and processing heuristics. Yet, the issue of how abstract formal competence might be overcome by heuristics (or vice versa?), and why, is left to speculation. The core abstract rules for natural language syntax are centered around the interface between recursive self-call, compositionality, and structure-dependence of element relations (Murphy 2025; Murphy et al. 2024). Moreover, to consider LLMs as models showing formal competence, LLMs would need to be able to take a firm stance in discerning *possible* from *impossible* natural language (Mitchell and Bowers, 2020). Their inability to do so (Figure 2.6) leaves one wondering about how the presumed rules/heuristics interplay would orchestrate so as to putatively allow LLMs to formally behave in a human-like manner. Mahowald et al. (2023) speak of artificial systems parsing "coherent stretches of text," but language is not a linear system of strings. It is purely an abstract hierarchical organizational system that may indeed interface with systems of externalization that demand seriality, but not necessarily (see Chapter 1 of this volume).

Figure 2.6: An LLM failure to distinguish between possible and impossible language.

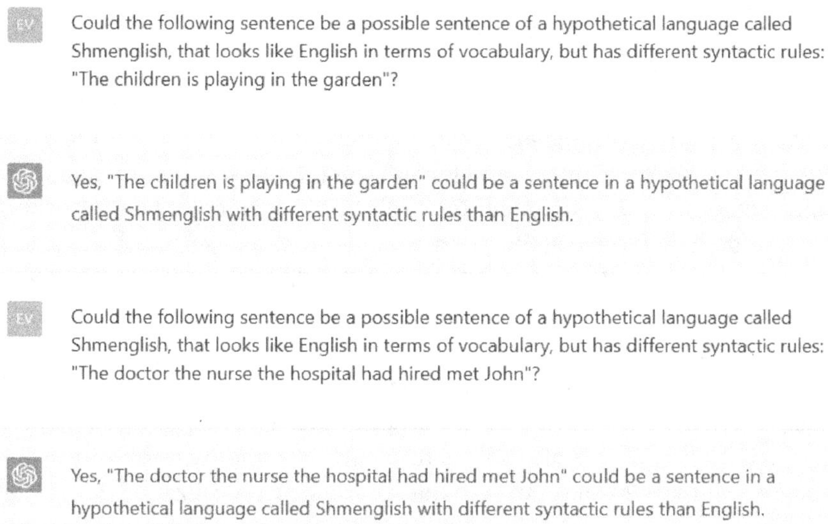

The first Shmenglish sentence (Shmenglish #1) is ungrammatical in Standard English, but it is grammatical in other varieties of English (e.g., Belfast English; Henry 2005). The second Shmenglish example (Shmenglish #2; Frazier 1985) is an impossible sentence. Drawing a distinction between absolute and relative ungrammaticality (Leivada and Westergaard 2020), Shmenglish #1 is ungrammatical in the relative sense: not forming part of a target grammar. Shmenglish #2 is different because it violates a core principle of language: Chomsky's (1981) θ-criterion, according to which each argument bears one and only one θ-role, and each θ-role is assigned to one and only one argument. As such, Shmenglish #2 is unlikely to ever form part of any grammar, and it is ungrammatical in the absolute sense (Leivada and Westergaard 2020). For a more extensive survey of grammatical abilities of OpenAI's advanced reasoning model o3-mini-high that provides converging evidence in this direction, see Murphy et al. (2025a).

Building on the juxtaposition of human and LLM formal competence, Mahowald et al. (2023) explore such a comparison at the neuroanatomical level. With reference to the neuroscientific evidence their main argument rests on, which partially leverages on dissociations in neuroatypical populations (e.g., the language/thought dissociation in global aphasia; Fedorenko and Varley, 2016), it is important to point out that the existence of neuroatypical dissociations is proof of integration of allegedly separate modules in neurotypical functioning. That is, the fact that formal and functional competence (a distinction that should itself be additionally explored and justified)[1] might synchronically recruit different neurocognitive mechanisms does not entail that language and thought come to ontogenetically evolve in individuals as two separate entities, which would amount to considering functional competence as utterly disregarding the remainder of cognition. There is, in fact, evidence that even frequency, possibly the simplest statistical property of words in a language, might at times act as a direct reflection of our perception of reality (Günther and Rinaldi, 2022, and references therein). This alone would make any argument for a robust dissociation of language "from the rest of high-level cognition, as well as from perception and action" (Mahowald et al. 2023: 6) untenable, alongside related arguments that draw a sharp line between formal linguistic competence and the remainder of cognition. Lastly, there are many aspects of functional cognition and thought that language might influence and regulate (in both the developing and adult brain) above and beyond the general complex reasoning tasks that are tested in various aphasias. Syntax also seems to tightly constrain types of expressible thought (Murphy 2020), and even the process of lexicalization re-formats concepts from core knowledge systems in ways that seem to imbue them with domain-specific representational features (Pietroski 2018; Pustejovsky and Batiukova 2019).

Overall, LLMs face important challenges in terms of both form and function. Despite claims about the mastery of formal competence, the information such models provide about the target system amounts to a distorted view of it. Given that target system-model discrepancies are yet to be measured (e.g., to what degree can LLMs tell apart possible from impossible languages in large-scale, systematic testing? Is the number of hallucination-featuring outputs constant across models and languages?), the absence of content transparency becomes a major concern.

5. Defining the user

If LLMs generated transparent and reliable data, they could be informative to scholars of different fields. Linguists, as scientists who investigate language, seem to be in an eminent position to examine the outputs of such models and

---

[1] Mahowald et al. (2023, 4) argue: "It has turned out that quite a lot of linguistic knowledge, e.g., about syntax and semantics, can be learned from language data alone [...]". In talking about semantics, the line between formal and functional competence becomes even blurrier, so that functional competence here can hardly be spared from discussion.

evaluate their predictive or explanatory power. At the same time, the research questions that guide the modelling need to be defined *first*. Finding an effect (for instance, a frequency effect in LLMs' ability to establish long-distance agreement) is a result that awaits explanation; it is not an explanation itself (van Rooij and Baggio 2021).

The gist of user-related issues boils down to the following question: With what purpose are LLMs developed and studied? One possible answer is to understand the properties of language. However, if we have already established that a mismatch exists between the target system and the content of the model, any conclusions drawn based on the model are potentially wrong. Consequently, any attempt to match the two (i.e., model and target system) is fraught with inherent interpretative difficulties.

The user-related challenges this situation raises are shown in Mahowald et al. (2023), who discuss Chomsky's view that deep learning tells us nothing about human language. After presenting Chomsky's view, Mahowald et al. cite a long line of position pieces that precisely connect LLMs with aspects of human language processing. The fact that a specific linguist does not view LLMs as informative about the questions they ask regarding human language cannot be taken to entail that the integration of the study of LLMs to the study of language "still encounters resistance," as Mahowald et al. (2023, 1) put it, any more than the reluctance of any prominent neurolinguist to address research questions from classical Indian grammatical theory can be taken to demonstrate general resistance to developing neural models of syntax. A field of study cannot be said to resist interaction with another potentially related field because some scholars working within it may find that a specific *tool* is not sufficiently informative *for their research questions*. In other words, what Mahowald et al. (2023) leave in the margins is the fact that the research questions need to be defined first, and then the utility of LLMs will be examined. In other words, Chomsky did not offer an opinion about whether someone interested in human language should approach their topic through the use of LLMs. He makes a much narrower claim, which is based on the absence of theoretically informed hypotheses about the precise linguistic questions that drive the development of LLMs. More specifically, in the 2022 *Debunking the great AI lie* Web Summit, Chomsky said the following, when asked whether he thinks that there has been any contribution from the study of LLMs to the actual understanding of linguistics to which he has devoted his entire career: "I can't think of one single thing. [...] Some days ago, a friend sent me an article from something called, I think, the Federation of Associations of Behavioral and

Computer Science, or something like that. It was a massive study in which the guy had all of Reddit and a ton of other things in his database, all kind of super computers, studying what you can learn about frequency of words [...]". In recent work (Chomsky 2022), he re-affirmed this position: AI "may be useful for linguistics, like maybe it can study huge number amounts of data and find properties that were not noticed, something like that. It could be a good tool" (p. 364). Chomsky, thus, recognizes that valuable work that pertains to some aspect of linguistics can be carried out through LLMs. However, a good tool is not *ipso facto* informative about the key research questions that guide Chomsky's (and many other linguists') research, outlined in (1)-(5):

1. What is knowledge of language?
2. How is that knowledge acquired?
3. How is that knowledge put to use?
4. How is that knowledge implemented in the brain?
5. How did that knowledge emerge in the species? (Boeckx and Grohmann 2007: 1, following Chomsky 1986)

Of course, different scholars may have a different conception of the *quo vadis* of linguistics in terms of its key research questions than Chomsky. Such a plurality of different views is typical in all fields of science, and attested even within Chomsky's theoretical niche. For instance, in *Me and Chomsky: Remarks from Someone Who Quit*, Sascha Felix writes about the nature and the orientation of current work in linguistics:

> "In some sense I feel that much (but obviously not all) of current linguistic work displays a relapse to the spirit prevailing in pre-Chomskyan times. *Linguistics is about describing language data. Period. Beyond this there is no deeper epistemological goal. Of course, those who became linguists because they like to play around with language data could not care less, because they can pursue their interests under any development of the field, nowadays possibly with less pressure and stress.* Personally, I felt that much of what I was offered to read in recent years was intolerably boring and that the field of linguistics was becoming increasingly uninteresting and trivialized." (2010, 71, emphasis added)

While describing patterns of word frequency is undoubtedly worth studying, Chomsky's argument is that using AI to do it will not help us shed new light on the core research questions in (1)-(5), even though it may still have a positive contribution to the world by enabling the development of useful tools that make our lives easier. One of the reasons has to do with the approach one has towards conducting scientific work. Some scholars believe that scientific research should not start with tools; it starts with theoretically informed hypotheses about specific research questions, which different tools can then help explore from different perspectives. While new technologies have the power to determine the questions that are being asked (Lewontin 2000), this simply opens a new circle of experimentation (Figure 2.7) that re-defines the starting hypotheses and formulates new research questions.

Figure 2.7: The circle of experimentation, taking language as the domain of study

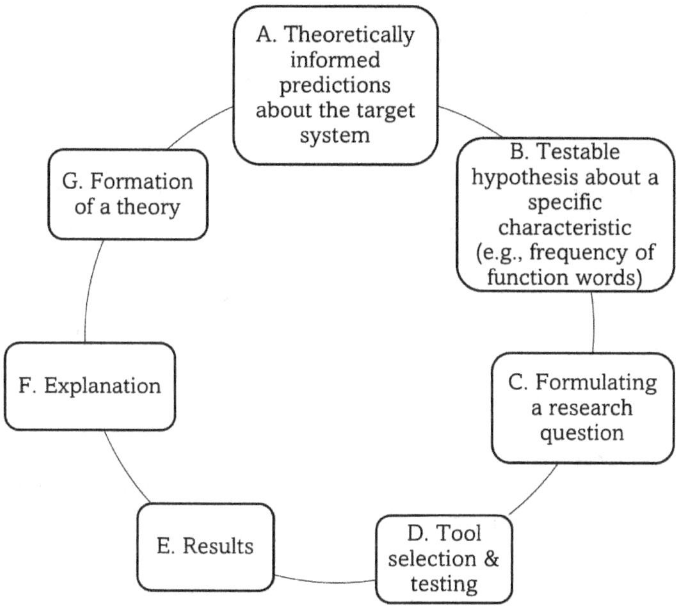

Employing this perspective, asking whether Chomsky (or anyone) believes that LLMs hold promise for the study of language puts things in an impoverished framework unless one first articulates specific research questions that LLMs aim to answer. The overall aim should be the discovery not of computationally or experimentally constructed effects through cutting-edge AI tools, but of plausible *explanations* of some concrete aspect of real-world abilities such as language (van Rooij and Baggio 2021). From this perspective, it is not clear what explanations LLMs can help adduce about the real-world properties of the target system, considering the risks identified here. To tame the AI hype in favor of scientific progress that tells us something meaningful about human language, we need (i) a more transparent representation of LLMs and their training sets, and (ii) a clear formulation of the research questions that guide their development and exploitation.

6. Outlook

Bill Gates heralded the arrival of GPT as "the most important advance in technology since the graphical user interface" (Gates 2023). We have argued here that only by taking the lessons of theoretical linguistics seriously will modern LLMs approximate human-like performance on language tasks. We also remain sceptical that LLMs will inform linguistic theory. LLMs at least need to be augmented with the appropriate inductive biases (e.g., syntactic priors). We discuss in this final section some prospects for, and limitations of, current research.

Consider first Transformer Grammars (Sartran et al. 2022), a syntactic language model implementing recursive syntactic composition of phrase representations through attention, and showing improved performance when internalizing structural properties of language, rather than abandoning the insights of theoretical linguistics. Importantly, Sartran et al. (2022) show that including an inductive bias that constrains the model to explain the data through built-in recursive syntactic composition operations ultimately aids with scalability, rather than hindering it.

More recently, the BabyLM Challenge attempts to approach human-like artificial language learning (Warstadt et al., 2023), but results are yet to show that LLMs that have been restricted to the same inputs as an infant can generalize the way humans do. Other work using ecologically valid datasets has highlighted even further the importance of syntactic priors. Yedetore et al. (2023) trained LSTMs and Transformers on text from the CHILDES corpus. While the models performed well at capturing the surface statistics of child-directed speech, they generalized in a way more consistent with an incorrect linear rule than the correct hierarchical rule for English Yes/No questions. As such, the general sequence-processing biases of standard neural network architectures are insufficient to capture core properties of English syntax. Other work that has embedded syntactic structure-sensitivity biases into neural networks has shown promising results that more closely mimic human judgements (McCoy et al. 2020), and that the cognitive plausibility of a corpus has clear effects on how it performs on syntactic processes (Huebner et al. 2021).

While it is certainly true that learners can infer hierarchical structure purely from positive evidence (Lan et al., 2022), it does not follow that this suffices for humans to establish form-meaning mappings for language. LLMs would become more relevant to linguistic theory if, for example, (i) an LLM predicted that sentence type A was more difficult to parse than type B; (ii) we have no independent theoretical reason to assume this to be true; (iii) and this prediction of the LLM was found to be true of human parsing. To our knowledge, this type of insight has yet to be forthcoming. In this connection, while the sequence of hidden state vectors and attention patterns provided by LLMs often correlates with some notions from syntactic theory, they do not yet allow us to make new predictions for language parsing. The inability to inspect trained states, due to commercial confidentiality, is also a major obstacle to calling what LLMs do 'linguistic theory.' After all, if LLMs learn Python (a phrase structure grammar) as well as they learn natural language (Veres 2022), why are we not concluding that LLMs are unveiling the secrets of Python?

Although GPT-4 is multi-modal, so far it has failed to show competence for multi-modal complex language. If a system has only inferred structure for text-based language (and not sign language), then it has not necessarily learned language. Human language is not tied to a specific modality. Relatedly, echoing some old concerns of Harnad (1990), ChatGPT has no sensory grounding and so regularly makes false claims about physical properties of individuals (e.g., famous philosophers like David Chalmers have been claimed to be bald). The idea that LLMs are good models of actual human cognition (Piantadosi 2023) is simply not substantiated at any level. The facility of LLMs for predicting human behavioral and/or neuroimaging responses does not lead to explanatory power (Rawski and Baumont 2023).

If the goal of a syntactician is to characterize the language acquisition device, then current LLMs fail. Alongside the above inductive biases for syntactic structure, we also need a story for how the featural content of functional lexical items is set, varying as this does across languages (e.g., Tense, Complementizer, Determiner, Aspect, Voice, Classifier, Mood features). Indeed, this may form the essence of what we mean when we talk of 'different languages.' Other properties that need to be accounted for, but have so far not been addressed in the AI literature, include the setting of Person features, which exhibit robust, non-trivial generalizations that do not seem to be accounted for via domain-general learning mechanisms. For example, the morphological composition of Person, its interaction with Number, its connection to space, and properties of its semantics and linearization all appear to be strong candidates for our knowledge of language (Harbour 2016). Other examples that fail with current LLMs include certain island constraints (Katzir 2023), but see Wilcox et al. (2022) for some recent evidence that autoregressive language models can show island sensitivity.

Further afield, aspects of linguistic semantics also need to be accounted for in learning models, such as the principle of conservativity for nominal quantifiers; when 'every,' 'some' and 'most' take two arguments, they only consider individuals that satisfy their first argument (Barwise and Cooper 1981). Lacking any mapping from structures to conceptual inferences or cognitive models, LLMs also show poor performance at segregating notions of likelihood (world-knowledge-based information, commonsense, etc.) from grammaticality (Katzir 2023).

Moving beyond basic architectural components relying on structure-dependence and compositionality, there are a number of robust findings from theoretical linguistics that demand a learning and implementational account:

- Agent asymmetry: Noun Phrases (NPs) bearing Agent roles are higher than NPs bearing other roles in the unmarked structure of the clause.
- Extended projections: Clauses and nominals consist of a verbal and nominal head, dominated by zero or more members of an ordered sequence of functional elements.
- Adverbial hierarchies: There are semantically defined classes of adverbs that appear in the same hierarchical order in all languages in which they exist overtly.
- Pronoun binding: The conditions on pronominal reference cannot be stated purely with linear order.
- Quantifier raising: The logical scope of natural language quantifiers (over individuals, times or situations/worlds) does not have to match their surface position. Quantifier scope is co-determined by structural factors (islands, clausal boundaries), logical properties of the quantifier (universal vs. existential) and the form of the quantificational expression (simple vs. complex indefinites).

Until the basic properties of syntax are captured by LLMs—or the semantic properties of basic adjectives, which are also currently out of reach (Liu et al. 2023)—we echo the suggestion in Veres (2022) that these systems should be referred to as 'corpus models' and not language models. 'Scale' is far from all that is needed; what is lacking is an ability of LLMs "to abstract their knowledge and experiences in order to make robust predictions, generalizations, and analogies; to reason compositionally and counterfactually; to actively intervene on the world in order to test hypotheses; and to explain one's understanding to others" (Mitchell and Krakauer 2023).

Bibliography

Barwise, Jon, and Robin Cooper. 1981. "Generalized Quantifiers and Natural Language." *Linguistics and Philosophy* 4: 159–219. https://doi.org/10.1007/BF00350139

Bender, Emily M., and Alexander Koller. 2020. "Climbing towards NLU: On Meaning, Form, and Understanding in the Age of Data." In *Proceedings of the 58th Annual Meeting of the Association for Computational Linguistics*, 5185–5198. https://doi.org/10.18653/v1/2020.acl-main.463

Bender, Emily M., Timnit Gebru, Angelina McMillan-Major, and Shmargaret Shmitchell. 2021. "On the Dangers of Stochastic Parrots: Can Language Models Be Too Big?" In *Proceedings of the 2021 ACM Conference on Fairness, Accountability, and Transparency (FAccT '21)*, 610–623. https://doi.org/10.1145/3442188.3445922

Birhane, Abeba, and Olivia Guest. 2021. "Towards Decolonising Computational Sciences." *Kvinder, Køn & Forskning* 29 (2): 60–73. https://doi.org/10.7146/kkf.v29i2.124899

Birhane, Abeba, Pratyusha Kalluri, Dallas Card, William Agnew, Ravit Dotan, and Michael Bao. 2022. "The Values Encoded in Machine Learning Research." In *Proceedings of the 2022 ACM Conference on Fairness, Accountability, and Transparency (FAccT '22)*, 173–184. https://doi.org/10.1145/3531146.3533083

Boeckx, Cedric, and Kleanthes K. Grohmann. 2007. "The Biolinguistics Manifesto." *Biolinguistics* 1: 1–8. https://doi.org/10.5964/bioling.8583

Chomsky, Noam. 1981. *Lectures on Government and Binding*. Dordrecht: Foris Publications.

Chomsky, Noam. 1986. *Knowledge of Language: Its Nature, Origin and Use*. New York: Praeger.

Chomsky, Noam. 2022. "Genuine Explanation and the Strong Minimalist Thesis." *Cognitive Semantics* 8 (3): 347–365. https://doi.org/10.1163/23526416-bja10040

Dentella, Vincenzo, Florian Günther, and Evelina Leivada. 2023. "Systematic Testing of Three Language Models Reveals Low Language Accuracy, Absence of Response Stability, and a Yes-Response Bias." *Proceedings of the National Academy of Sciences* 120 (51): e2309583120. https://doi.org/10.1073/pnas.2309583120

Dentella, Vincenzo, Elliot Murphy, Gary Marcus, and Evelina Leivada. 2023. "Testing AI Performance on Less Frequent Aspects of Language Reveals Insensitivity to Underlying Meaning." *arXiv* preprint, arXiv:2302.12313

Fedorenko, Evelina, and Rosemary Varley. 2016. "Language and Thought Are Not the Same Thing: Evidence from Neuroimaging and Neurological Patients." *Annals of the New York Academy of Sciences* 1369 (1): 132–53. https://doi.org/10.1111/nyas.13046

Felix, Sascha W. 2010. "Me and Chomsky: Remarks from Someone Who Quit." In *Language and Logos: Studies in Theoretical and Computational Linguistics. Festschrift for Peter Staudacher on His 70th Birthday*, edited by Thomas Hanneforth and Gisbert Fanselow, 64–71. Berlin: Akademie Verlag. https://doi.org/10.1524/9783050062365.64

Frazier, Lyn. 1985. "Syntactic Complexity." In *Natural Language Parsing: Psychological, Computational and Theoretical Perspectives*, edited by David R. Dowty, Lauri Karttunen, and Arnold Zwicky, 129–189. Cambridge: Cambridge University Press. https://doi.org/10.1017/CBO9780511597855.005

Frigg, Roman, and Stephan Hartmann. 2020. "Models in Science." In *The Stanford Encyclopedia of Philosophy*, edited by Edward N. Zalta. Stanford, CA: Metaphysics Research Lab, Stanford University. https://plato.stanford.edu/archives/spr2020/entries/models-science/

Gates, Bill. 2023. "The Age of AI Has Begun." *GatesNotes*, March 21.

Guest, Olivia, and Andrea E. Martin. 2021. "How Computational Modeling Can Force Theory Building in Psychological Science." *Perspectives on Psychological Science* 16 (4): 789–802. https://doi.org/10.1177/1745691620970585

Günther, Florian, and Laura Rinaldi. 2022. "Language Statistics as a Window into Mental Representations." *Scientific Reports* 12: 8043. https://doi.org/10.1038/s41598-022-12027-5

Harbour, Daniel. 2016. *Impossible Persons*. Cambridge, MA: MIT Press. https://doi.org/10.7551/mitpress/9780262034739.001.0001

Harnad, Stevan. 1990. "The Symbol Grounding Problem." *Physica D: Nonlinear Phenomena* 42 (1–3): 335–346. https://doi.org/10.1016/0167-2789(90)90087-6

Henry, Alison. 2005. "Non-Standard Dialects and Linguistic Data." *Lingua* 115: 1599–1617. https://doi.org/10.1016/j.lingua.2004.07.006

Huebner, Philip A., Eyal Sulem, Cynthia Fisher, and Dan Roth. 2021. "BabyBERTa: Learning More Grammar with Small-Scale Child-Directed Language." In *Proceedings of the 25th Conference on Computational Natural Language Learning*, 624–646. https://doi.org/10.18653/v1/2021.conll-1.49

Katzir, Roni. 2023. "Why Large Language Models Are Poor Theories of Human Linguistic Cognition: A Reply to Piantadosi." *Biolinguistics* 17: e13153. https://doi.org/10.5964/bioling.13153

Lan, Nan, Marc Geyer, Emmanuel Chemla, and Roni Katzir. 2022. "Minimum Description Length Recurrent Neural Networks." *arXiv* preprint, arXiv:2111.00600. https://doi.org/10.1162/tacl_a_00489

Leivada, Evelina, Elliot Murphy, and Gary Marcus. 2023. "DALL·E 2 Fails to Reliably Capture Common Syntactic Processes." *Social Sciences & Humanities Open* 8: 100648. https://doi.org/10.1016/j.ssaho.2023.100648

Leivada, Evelina, and Marit Westergaard. 2020. "Acceptable Ungrammatical Sentences, Unacceptable Grammatical Sentences, and the Role of the Cognitive Parser." *Frontiers in Psychology* 11: 364. https://doi.org/10.3389/fpsyg.2020.00364

Lemoine, Blake. 2022. "Is LaMDA Sentient?—An Interview." *Medium*, June 11. https://cajundiscordian.medium.com/is-lamda-sentient-an-interview-ea64d916d917

Lewontin, Richard. 2000. *The Triple Helix: Gene, Organism, and Environment.* Cambridge, MA: Harvard University Press.

Liu, Wei, Ming Xiang, and Nai Ding. 2023. "Adjective Scale Probe: Can Language Models Encode Formal Semantics Information?" Manuscript, Zhejiang University. https://doi.org/10.1609/aaai.v37i11.26559

Mahowald, Kyle, Anna A. Ivanova, Idan A. Blank, Nancy Kanwisher, Joshua B. Tenenbaum, and Evelina Fedorenko. 2023. "Dissociating Language and Thought in Large Language Models: A Cognitive Perspective." *arXiv* preprint, arXiv:2301.06627. https://arxiv.org/abs/2301.06627

Marcus, Gary, Ernest Davis, and Scott Aaronson. 2022. "A Very Preliminary Analysis of DALL·E 2." *arXiv* preprint, arXiv:2204.13807. https://arxiv.org/abs/2204.13807

Marcus, Gary, and Ernest Davis. 2019. *Rebooting AI: Building Artificial Intelligence We Can Trust.* New York: Pantheon Books.

McCoy, R. Thomas, Robert Frank, and Tal Linzen. 2020. "Does Syntax Need to Grow on Trees? Sources of Hierarchical Inductive Bias in Sequence-to-Sequence Networks." *Transactions of the Association for Computational Linguistics* 8: 125–140. https://doi.org/10.1162/tacl_a_00304

Mitchell, James, and Jeffrey S. Bowers. 2020. "Priorless Recurrent Networks Learn Curiously." In *Proceedings of the 28th International Conference on Computational Linguistics*, 5147–5158. https://doi.org/10.18653/v1/2020.coling-main.451

Mitchell, Melanie. 2019. *Artificial Intelligence: A Guide for Thinking Humans.* New York: Farrar, Straus and Giroux.

Mitchell, Melanie, and David C. Krakauer. 2023. "The Debate over Understanding in AI's Large Language Models." *Proceedings of the National Academy of Sciences of the United States of America* 120 (13): e2215907120. https://doi.org/10.1073/pnas.2215907120

Murphy, Elliot. 2020. "Language Design and Communicative Competence: The Minimalist Perspective." *Glossa* 5 (1): 2. https://doi.org/10.5334/gjgl.1081

Murphy, Elliot. 2025. "ROSE: A Universal Neural Grammar." *Cognitive Neuroscience.* https://doi.org/10.1080/17588928.2025.2523875

Murphy, Elliot, Evelina Holmes, and Karl Friston. 2024. "Natural Language Syntax Complies with the Free-Energy Principle." *Synthese* 203: 154. https://doi.org/10.1007/s11229-024-04566-3

Murphy, Elliot, Evelina Leivada, Vincenzo Dentella, Florian Günther, and Gary Marcus. 2025. "Fundamental Principles of Linguistic Structure Are Not

Represented by o3." *arXiv* preprint, arXiv:2502.10934. https://doi.org/10.5964/bioling.19021

Murphy, Elliot, Peter S. Rollo, Katrien Segaert, Peter Hagoort, and Nitin Tandon. 2024. "Multiple Dimensions of Syntactic Structure Are Resolved Earliest in Posterior Temporal Cortex." *Progress in Neurobiology* 241: 102669. https://doi.org/10.1016/j.pneurobio.2024.102669

Murphy, Elliot, Oscar Woolnough, Peter S. Rollo, Zachary Roccaforte, Katrien Segaert, Peter Hagoort, and Nitin Tandon. 2022. "Minimal Phrase Composition Revealed by Intracranial Recordings." *Journal of Neuroscience* 42 (15): 3216–3227. https://doi.org/10.1523/JNEUROSCI.1575-21.2022

Murphy, Elliot, Jill de Villiers, and Sofia L. Morales. 2025. "A Comparative Investigation of Compositional Syntax and Semantics in DALL·E and Young Children." *Social Sciences & Humanities Open* 11: 101332. https://doi.org/10.1016/j.ssaho.2025.101332

Pagnoni, Artidoro, Vidhisha Balachandran, and Yulia Tsvetkov. 2022. "Understanding Factuality in Abstractive Summarization with FRANK: A Benchmark for Factuality Metrics." In *Proceedings of the 2022 Conference of the North American Chapter of the Association for Computational Linguistics (NAACL)*, 4812–4829. https://doi.org/10.18653/v1/2021.naacl-main.383

Piantadosi, Steven T. 2023. "Modern Language Models Refute Chomsky's Approach to Language." Manuscript. https://ling.auf.net/lingbuzz/007180

Piantadosi, Steven T., and Felix Hill. 2022. "Meaning without Reference in Large Language Models." In *Proceedings of the 36th Conference on Neural Information Processing Systems (NeurIPS)*.

Pietroski, Paul M. 2018. *Conjoining Meanings: Semantics without Truth Values.* Oxford: Oxford University Press. https://doi.org/10.1093/oso/9780198812722.001.0001

Potts, Christopher. 2020. "Is It Possible for Language Models to Achieve Language Understanding?" *Medium*, September 21. https://chrisgpotts.medium.com/is-it-possible-for-language-models-to-achieve-language-understanding-81df45082ee2

Pustejovsky, James, and Olga Batiukova. 2019. *The Lexicon.* Cambridge: Cambridge University Press. https://doi.org/10.1017/9780511982378

Rassin, Roee, Shauli Ravfogel, and Yoav Goldberg. 2022. "DALLE-2 Is Seeing Double: Flaws in Word-to-Concept Mapping in Text-to-Image Models." *arXiv* preprint, arXiv:2210.10606. https://arxiv.org/abs/2210.10606

Rawski, Jonathan, and Lucien Baumont. 2023. "Modern Language Models Refute Nothing." Manuscript. https://lingbuzz.net/lingbuzz/007203

Sartran, Laurent, Samuel Barrett, Adhiguna Kuncoro, Miloš Stanojević, Phil Blunsom, and Chris Dyer. 2022. "Transformer Grammars: Augmenting Transformer Language Models with Syntactic Inductive Biases at Scale." *Transactions of the Association for Computational Linguistics* 10: 1423–1439. https://doi.org/10.1162/tacl_a_00526

Schrimpf, Martin, Idan Asher Blank, Greta Tuckute, Cameron Kauf, Eghbal A. Hosseini, Nancy Kanwisher, Joshua B. Tenenbaum, and Evelina Fedorenko. 2021. "The Neural Architecture of Language: Integrative Modeling Converges on Predictive Processing." *Proceedings of the National Academy of Sciences* 118 (45): e2105646118. https://doi.org/10.1073/pnas.2105646118

Sprouse, Jon, and Diego Almeida. 2012. "Assessing the Reliability of Textbook Data in Syntax." *Journal of Linguistics* 48: 609–652. https://doi.org/10.1017/S0022226712000011

Sprouse, Jon, Carson T. Schütze, and Diego Almeida. 2013. "Assessing the Reliability of Journal Data in Syntax: *Linguistic Inquiry* 2001–2010." *Lingua* 134: 219–248. https://doi.org/10.1016/j.lingua.2013.07.002

van Rooij, Iris. 2022. "Psychological Models and Their Distractors." *Nature Reviews Psychology* 3: 127–128. https://doi.org/10.1038/s44159-022-00031-5

van Rooij, Iris, and Giosuè Baggio. 2021. "Theory before the Test: How to Build High-Verisimilitude Explanatory Theories in Psychological Science." *Perspectives on Psychological Science* 16 (4): 682–697. https://doi.org/10.1177/1745691620970604

Tam, Derek, André Mascarenhas, Shiyue Zhang, Stanley Kwan, Mohit Bansal, and Colin Raffel. 2022. "Evaluating the Factual Consistency of Large Language Models through Summarization." *arXiv* preprint, arXiv:2211.08412. https://doi.org/10.18653/v1/2023.findings-acl.322

Veres, Calin. 2022. "Large Language Models Are Not Models of Natural Language: They Are Corpus Models." *arXiv* preprint, arXiv:2112.07055. https://doi.org/10.1109/ACCESS.2022.3182505

Warstadt, Alex, Leshem Choshen, Aaron Mueller, Adina Williams, Ethan Wilcox, and Chengxu Zhuang. 2023. "The BabyLM Challenge: Sample-Efficient Pretraining on a Developmentally Plausible Corpus." *arXiv* preprint, arXiv:2301.11796. https://doi.org/10.18653/v1/2023.conll-babylm.1

Weidinger, Laura, John Uesato, Marcin Rauh, et al. 2022. "Taxonomy of Risks Posed by Language Models." In *Proceedings of the 2022 ACM Conference on Fairness, Accountability, and Transparency (FAccT '22)*, 214–229. https://doi.org/10.1145/3531146.3533088

Wilcox, Ethan G., Richard Futrell, and Roger Levy. 2022. "Using Computational Models to Test Syntactic Learnability." *Linguistic Inquiry* 55 (4): 805–848. https://doi.org/10.1162/ling_a_00491

Yedetore, Anhad, Tal Linzen, Robert Frank, and R. Thomas McCoy. 2023. "How Poor Is the Stimulus? Evaluating Hierarchical Generalization in Neural Networks Trained on Child-Directed Speech." *arXiv* preprint, arXiv:2301.11462. https://doi.org/10.18653/v1/2023.acl-long.521

Chapter 3

Language models as function approximators of text data: disrupting comprehension through and adversarial attack

David J. Lobina

Rovira i Virgili University

[A] computer program that succeeded in generating sentences of a language would be, in itself, of no scientific interest unless it also shed some light on the kinds of structural features that distinguish languages from arbitrary, recursively enumerable sets.

Chomsky (1963, 360)

Abstract: Large Language Models such as ChatGPT are examples of Machine Learning models (ML), and as such, they are best regarded as "function approximators," or vast correlation machines. That is, given a dataset, however big, of pairs of inputs and outputs, a ML network will construct a function that best describes (approximates) the distribution in the data, and from whence it can generate similar input-output pairs if prompted. In the case of ChatGPT, the dataset is composed of text—the products of the language faculty. In this sense, ChatGPT does not model a natural language per se, but the statistical distribution of text, the latter treated as recorded, static data. *A fortiori*, the claim that some of the results emanating from the development of Large Language Models bear any consequence for the study of language, especially regarding linguistic competence and acquisitional studies, appear to be the case of committing a category mistake.

Keywords: Knowledge of language, Linguistic Theory, Large Language Models, Artificial Intelligence

***

1. What can large language models be theories of?

We have been here before. In the 1980s and 90s, connectionist networks were all the rage in cognitive science, touted as they then were as possible models of cognition, in contrast to what is often termed as the 'standard theory' of cognition (Samuels 2010), an account of cognition in terms of symbolic representations and computations over the structural (thus syntactic) properties of these representations. Connectionism offered a different theoretical perspective, one based on algorithms tracking the distributional statistics of the input data the model was fed, with such data represented in the form of vectors and matrices (thus forgoing the use of explicit syntactic representations) and the algorithms implemented in a neural network architecture of nodes and edges designed to "learn" patterns from the inputted data (hence, the term "machine learning"). The successes of connectionist networks at the time were somewhat limited in scope and often not entirely applicable to the study of cognition, and their prominence in cognitive science diminished a bit in the 2000s.

The interest in this kind of model is now back in force since the advent of Large Language Models (LLMs), a species of machine learning with a slightly different architecture to connectionist networks (viz., the so-called transformers) and an improved processing model (parallel processing rather than sequential). The results of LLMs, at first sight at least, are clearly impressive and have revitalized the field of Artificial Intelligence (AI), in both academia and the private sector. The interest within cognitive science has also been significant, especially as it pertains to the study of language (see, e.g., Contreras Kallens et al. 2023). As was the case with connectionist networks, however, there is a significant disconnect between what these models produce—a given output—and the nature of the models that yield such outputs—and this, I shall argue here, dims the significance of LLMs to the study of language and cognition. This chapter will, therefore, engage the question of whether LLMs are, or can be regarded as, models of human cognition, and in particular as theories of language (including language processing and learning/acquisition).

In order to do so, this section outlines the main features of LLMs, briefly compares LLMs to the aforementioned standard account of cognition, and, in turn, to linguistic theories of language, with the aim of highlighting the shortcomings of LLMs as models of natural language on principled grounds. Building on this, the following section will delineate how so-called "adversarial attacks" can disrupt the operations of LLMs, thereby producing outputs that belie two important facts about these systems: LLMs model text data rather than a natural language (or the language capacity), and such text data are modelled in terms of mathematical properties assigned to them by the models, and not as linguistic outputs per se. In turn, and following up from this, the concluding section will emphasize that LLMs are "function approximators," effectively dealing with numbers and numbers only—they are not models of language or cognition.

AI is often understood as a computational system that exhibits intelligent behavior—for instance, complex behavior conducive to reaching specific goals. So understood, classical AI models from the 1940s or so, and until the so-called AI Winter in the 1970s, for the most part involved symbol-manipulation systems, and their alignment to cognitive science was explicitly recognized in these terms (Pylyshyn 1984). Further, such systems employed many of the constructs commonly found in formal logic and were often inspired by the theories of human cognition then dominant—among others, the aforementioned representational and computational paradigm, one version of which is back en vogue now (see Quilty-Dunn et al. 2023, including the responses the article received in the same issue of the journal).

There are, as a matter of fact, various approaches to AI, from logic- and knowledge-based ones to purely statistical methods involving Bayesian inference. Utterly dominant today is the use of machine learning algorithms, and in particular, deep learning methods. Indeed, by AI these days, it is the artificial neural networks of such approaches that most people have in mind, and hardly any of the other approaches shown in Figure 3.1.

Loosely based on the neurons of biological brains, though not thereby constituting a theory of how the brain works in any way, and closely related to statistics and probability theory to boot, the aim of machine learning models is to find generalizable predictive patterns in a dataset. Less roughly, machine learning involves creating a model that has been trained on a dataset (the so-called training data), and from this trained model, the algorithm attempts to process additional data in order to make predictions in the form of new outputs.

Figure 3.1: The Different Kinds of Artificial Intelligence

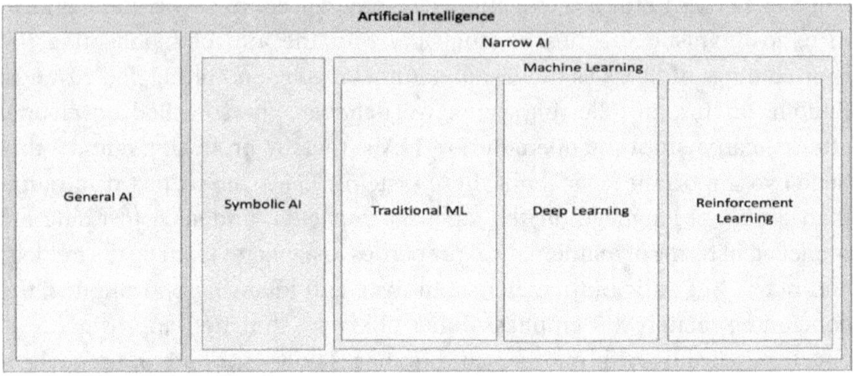

It is in this vein that neural networks (in short, NNs) can be said to "learn" patterns from the examples the networks are fed. The way this is set up is by establishing "communication" between populations of neurons (or nodes), which is to say by the transmission of actual numbers among nodes, as this is a mathematical model at heart. The output, that is to say, the intended outcome of the overall process (in effect, a set of neurons), is computed by some non-linear function of the sum of its inputs, and if there are multiple hidden layers mediating between the input and output neurons, then we are dealing with a so-called deep learning model, as shown in Figure 3.2.

Figure 3.2: Rough outline of a deep neural network

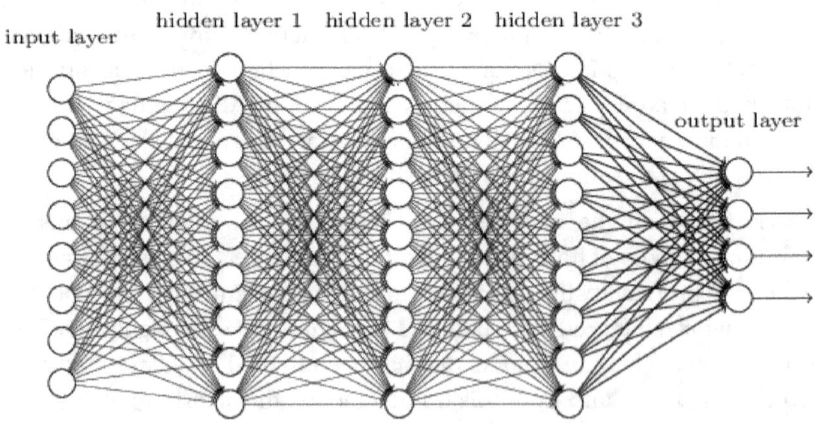

In effect, NNs track the various correlations between input and output layers that the model finds in the training dataset, and such "pattern finding" algorithms have proven to be very useful in all sorts of tasks, as in image classification, and now, language generation. What the approach amounts to is a situation in which the input neurons represent, say, images of pandas, and the output neurons are expected to classify the input images as images of pandas; in the case of LLMs, input neurons typically represent an incomplete string of words, and the output neurons will generate the probable next word for that string.

This picture is certainly applicable to LLMs. These models generate strings of words (more accurately, of word tokens) based on combinations from the training dataset, and do so by calculating so-called embeddings and binding relations in the data (no need to delve into these details here, but see below), the predictions ultimately grounded on something a human once produced.

To be more accurate, LLMs are large networks of matrix/tensor products, they calculate the probability of the next word given a context (the string), and they do so by representing words as vectors of values from which to calculate the probability of each word (via a logistic regression and some other computational processes), with sentences also represented as vectors of values. As alluded to earlier, most modern LLMs use transformers, which allow the models to carry out matrix calculations over these vectors, and the more transformers are employed (so-called attention layers), the more accurate the predictions appear to be.

At first sight, then, machine learning would appear to bear little resemblance to the manner in which humans learn language or cognition in general develops; language models compute input-output correlations on the data, and cognition is not quite like that. Indeed, it is important to note that even in those cases in which intelligent behavior seems to be displayed by a machine learning network, all that the system is effecting is a correlation between an input and an output—that is, it finds statistical patterns in the data—with no explicit representation of underlying causes (or understanding of language), and often manifesting no transparency as how the correlations were calculated in the first place. If so, it would appear to be a mistake to ascribe to LLMs any kind of equivalence to basic features of human cognition, as LLMs do not employ the right cognitive mechanisms to that end.

This point alludes to a distinction drawn by Pylyshyn (1984) between the weak equivalence of two computational systems and their strong equivalence. Weak equivalence obtains when two systems compute the same input-output function (thus producing the same *function in extension*, in the sense of Church 1941; see Lobina 2017 for a discussion of how this applies to cognition), whilst strong equivalence obtains when two computational systems go beyond this and accomplish weak equivalence by executing the same algorithm in the same architecture (thereby producing the same *function in intension*; vide idem). Pylyshyn (1984) insisted that cognitive scientists were in the business of establishing the strong equivalence between their postulated models and actual properties of cognition, and according to this desiderata, what machine learning models demonstrate would be weak equivalence only.

At this point, it will be conducive to describe, albeit briefly, what the standard theory of cognition is meant to be in order to draw the right contrast to LLMs, and a good point to start is Quilty-Dunn et al. (2023), mentioned above. In arguing for a *language of thought* (LoT) architecture of cognition,[1] Quilty-Dunn et al. focus on data from three sources: certain aspects of early vision, the cognitive abilities of nonverbal minds (namely, reasoning with physical objects in preverbal infants and nonhuman animal species), and so-called System 1 cognition (namely, the use of heuristics in reasoning). Quilty-Dunn et al. argue that some, or perhaps even most, psychological processes involve operations over symbolic representations that are only sensitive to the syntactic properties of these representations, and to some extent, the paper does so by contrasting LLMs to the very properties of the LoT Quilty-Dunn et al. emphasize throughout the article—to wit, constituent structure, predicate-argument structure, logical operators, etc.

These features of cognition are most obvious when evaluating systematicity, a central ability of human cognition. This is the capacity to process and represent structurally related phenomena and is best described as a hierarchy of abilities. The following list is from Hadley (2004), where systematicity is related to linguistic abilities and ranked in increasing order of sophistication (the point is meant to generalize to cognition *tout court*, though):

---

[1] The standard theory of cognition is often termed the "language of thought hypothesis," but I should note that the phrase "the language of thought" can refer to two related but slightly different ideas, and one need not hold both views. One idea is the claim that most of our thinking does not take place in a natural language but in a conceptual representational system. The other, more controversial idea is the position Quilty-Dunn et al. are most concerned with – the claim that psychological processes involve computational operations over symbolic representations.

- Weak systematicity: an agent can process test sentences containing novel combinations of words.
- Quasi systematicity: this involves weak systematicity, and it is extended to embedded sentences—a sentence inside another sentence, as in *the rat [the cat bit] ran away.*
- Strong systematicity: weak systematicity plus the ability to process known words in novel positions (i.e., positions not present in the training set)—for example, this would be the ability to transform an active sentence, composed of known words, into a passive sentence, constituting a new pattern.
- Strong semantic systematicity: strong systematicity plus the ability to assign appropriate meanings to all words in any novel test sentence—for example, the ability to understand verbs encountered in the indicative (e.g., *John sings*) but now used in the imperative (*Sing, John!*).

Crucially, systematicity revolves around two important principles of cognition, according to proponents of an LoT account:

- Structure Sensitivity: complex representations have an internal structure, as they are composed of constituents that are literally there.
- Compositionality: "the content of thoughts is determined [...] by the content of the [...] concepts that are their constituents" (Fodor and McLaughlin 1990, 185).

As Hadley (2004) shows, the language models at the time (as mentioned, connectionist networks) did not fare very well, only demonstrating limited mastery of systematicity, and with rather simple sentences (the best networks were hybrid models, in fact, the combination of deep learning and symbolic representations, but this is by the by now). Do modern language models, equipped with transformers, do any better?

In a recent paper that does not mention systematicity explicitly but describes an ability that is close to it (viz., Berglund et al. 2024), it is found that language models suffer from what the authors call 'the reversal curse'—that is, the inability to generalize *B is A* when the model has been trained on *A is B*. Or less cryptically, language models are not able to answer questions such as 'who is the ninth Chancellor of Germany?' after being trained on sentences such as 'Olaf Scholz is the ninth Chancellor of Germany.' Consider the following pair of questions from Berglund et al. (2024) to clarify further:

(1) *Who is Tom Cruise's mother?*

(2) *Who is Mary Lee Pfeiffer's son?*

The model answers the question in (1) correctly 79% of the time, and unsurprisingly so, given that the relevant information was part of the training dataset (some LLMs are fed the entirety of the internet). Regarding the question in (2), however, the model answers it correctly only 33% of the time, even though the answer to this question is Tom Cruise and the model had been trained on (1) beforehand.

To be more specific, in this study, LLMs were trained on sentences of the form *<name> is <description>* and were then asked to predict the relevant piece of text in the opposite direction—that is, *<description> is <name>*. The ability to go from one form to the other is clearly related to the structure-sensitive ability of systematicity—and, in particular, to the availability of an argument-predicate structure, where the verb *is* constitutes the predicate and *names* and *descriptions* are the arguments.[2]

I am constantly emphasizing the point that what LLMs do is predict the next word given a string of words as a context—next token prediction, that is—but this is certainly not what a user of such systems experiences when dealing with something like ChatGPT. This is because ChatGPT is a "dialogue management system," an AI assistant or chatbot that queries an underlying LLM (the latest version at time of writing is GPT-5.2) so that the commands and questions the user poses elicit (or "prompt") the right output from the LLM, which in every case will be a single word. An LLM is not a chatbot, but a mathematical model of a dataset with a set of trained "parameters" (the sought-after weights of the aforementioned vectors and matrices). As the computer scientist Murray Shanahan (2023) has put it, when one asks an LLM a question such as "who was the first person to walk on the moon?", what the model is really being asked is something along the lines of:

> Given the statistical distribution of words in the vast public corpus of (English) text, what words are most likely to follow the sequence *The first person to walk on the Moon was*?

---

[2] Lake and Baroni (2023) have tackled the issue of systematicity within LLMs head-on and claim that a meta-learning model can, in fact, account for some systematic abilities. This study is focused on a much narrower version of systematicity than what I have outlined here, though, and it is unclear whether these models generalize to the most sophisticated examples of systematicity; moreover, the results appear to be modest even within their own models (see Goodale and Mascarenhas 2023).

In reality, an LLM is embedded within a larger system that includes background "prefixes" to coax the system into producing conversation-like behavior (e.g., a template of what a conversation looks like), but the LLM itself simply generates sequences of words that are statistically likely to follow from a specific prompt. It is through the use of prompt prefixes that LLMs can be directed to "perform" various tasks beyond dialoguing and behave as if they were reasoning or participating in a psycholinguistic experiment, even providing grammatical judgements on specific strings (I will come back to this). But the model itself remains a sequence predictor: it outputs and manipulates strings of words and it has no explicit understanding of what these strings mean.

The latter two points are not unrelated; it is well-established within linguistics that word strings do not exhibit the right format to represent meaning unambiguously, this being one of the reasons most linguists aim to account for the so-called "strong generative capacity" of language in their theories (the generation of structures), as there is a close relationship between meaning and structure in language—a significant part of meaning is typically read off the syntax of a phrase (which is then enriched through pragmatic processes). This is in contrast to the "weak generative capacity", namely, the generation of strings of elements, a construct that has proven to be more important to scholars working at the interface between language and formal systems—in fields such as computational linguistics and formal language theory, and now of course in AI in the form of language models.[3]

In the next section, I will delve deeper into what all this means for the purposes of evaluating LLMs as theories of language. I will start by discussing how linguistic knowledge is put to use during language comprehension and language acquisition, and then I will introduce the notion of an "adversarial attack" against machine learning models. Adversarial attacks will allow me to show how meaning considerations are not operative in LLMs in the way that they are in natural language, and this will offer an answer to the question posed as the title to this section; namely, that LLMs are not, and cannot be, theories

---

[3] See Dupre (2024) for a recent restatement of the now-classic point that the generation of strings does not capture the psychological reality of mentally represented grammars and their acquisition, an argument I will use myself in some form in the next section. The strong and weak generative capacities, incidentally, are related but distinct concepts to Pylyshyn's strong and weak equivalences; see Lobina (2017) for a more thorough discussion of the connection.

of language, and they also cannot serve as theories of linguistic production, be this in speech or text.

2. LLMs and the loss of comprehension: The case of adversarial attacks

Needless to say, there has been plenty of discussion in the literature regarding the very issues I have raised in the previous section. Contreras Kallens et al. (2023), for instance, have argued that LLMs constitute an "existence proof" that the ability to produce grammatical language can be learned from exposure to data alone, without the need to postulate language-specific processes or even representations. Katzir (2023) has instead argued that LLMs cannot be said to have learned a natural language, given that when evaluated, these models fail to detect some of the grammatical nuances that 12-year-olds and adults are unambiguously and effortlessly sensitive to (Dentella et al. 2023 provide a more thorough empirical study of the outputs of LLMs, drawing a similar conclusion).[4]

I would argue that we cannot conclude, from the apparent fact that LLMs correlate with human behavior, that such models are *ipso facto* theories of human capacities. My actual point here, though, is slightly different: LLMs have not learned an actual natural language, but information about the products of the language faculty, and in particular, *certain* information from text data only. And moreover, the language faculty cannot be equated to speech, text, or any other forms of externalization (see also Chapter 1 of this volume).

A common way to define a natural language in linguistics, and one that can be traced back to Ancient Greece (Panaccio 2017), is as a system that connects sound (or other externalizations of language, e.g., hand gestures, signs, written text) and meaning—a pair of meaning and articulation, if you will—the connection between the two mediated by syntax. And, in turn, a common way to define the underlying language faculty is as a mental system composed of a lexicon and a grammar, the latter a construct that includes various components, most centrally, syntax, semantics, and phonetics and phonology (and others).

Even though many scholars only have syntax in mind when talking of a "grammar," it should be stressed that for a given sentence to be grammatical,

---

[4] I would not want to exaggerate the significance of these "grammatical exercises" with LLMs, however. After all, asking a chatbot about the meaning of this or that sentence, or for grammatical judgements, is in a way quite absurd, for the simple reason that neither the chatbots nor the underlying LLMs have the right architecture to engage with such questions.

all the components I have mentioned need to align. This is true of any linguistic theory worth its salt, regardless of differences in focus and commitments. Consider two broad families of theories. For some theories, syntax sends structures to phonology and semantics for further operations, and in this sense the syntactic engine is said to be constrained by various conditions that phonology and semantics impose (see, for instance, Chomsky 2011), whilst for other theories syntax, semantics, and phonology operate in parallel, with various linking points during a derivation (see Jackendoff 2002). In any case, the overall picture yields important repercussions for the study of language.

Consider some of the processes involved in how children acquire language, for instance (I'm basing this material on Guasti 2002; cf. Dupre 2024). Children are sensitive to various phonological properties of the language or languages they are exposed to, basically from birth, and some of these features, such as intonation and stress, can help them identify categories such as determiners, work out where words start and end (in combination with some of the statistical patterns children can track), and even outline specific syntactic phrases (intonation can provide strong clues regarding when a syntactic phrase ends). Further, so-called semantic bootstrapping, according to which children can use semantic categories such as *objects* and *actions* as cues to identify syntactic categories such as *nouns* and *verbs*, respectively, can help children acquire the syntax of their language. And, conversely, so-called syntactic bootstrapping, which has it that children use "syntactic frames" to learn the meanings of words, can help children acquire the semantics of their language (for instance, the argument structure of a verb like *to ask* requires an *agent* doing the asking and a *topic*, or *theme*, to be asked about).

An intricate and systematic phenomenon, the result is an intricate and systematic knowledge of language, and this state of affairs also applies, necessarily, to language comprehension, where intonation can help hearers resolve potential ambiguities in the input they receive during verbal communication, the meanings of words can activate potential follow-ups, canonical word orders can help predict what meaning is being communicated, *e così via* (see, for instance, Fodor 1998a, 1998b).

Given these facts about natural language, it is not a little misguided to claim that LLMs produce "grammatical" language. If a given account of language does not cover the central role meaning has in how we acquire and process a natural language, along with shedding some light on the structure of language, then such an account will not be adequate. A fortiori, if linguistic meaning plays little to no role in your theory of language, then you do not have a theory of language at all.

In the case of LLMs, this point is more forcefully made by considering so-called adversarial attacks, a technique that attempts to "fool" neural networks by using a defective input. The following discussion will show that "meaning" receives no dispensation in an LLM, and this tells against LLMs being considered as theories of language comprehension (I shall base the following material partly on Xu et al. 2023; Fu et al. 2024; Zou et al. 2023).

An adversarial attack is an imperceptible perturbation to the original sample or input data of a machine learning model with the intention to disrupt its operations. Originally devised with machine vision models in mind, in these cases, the technique involves adding a small amount of noise to an image; in the case of Figure 3.3, it is the image of a panda, with the effect that the model misclassifies the image as that of a gibbon, and with an extremely high degree of confidence.

Figure 3.3: Adversarial attack on a visual input

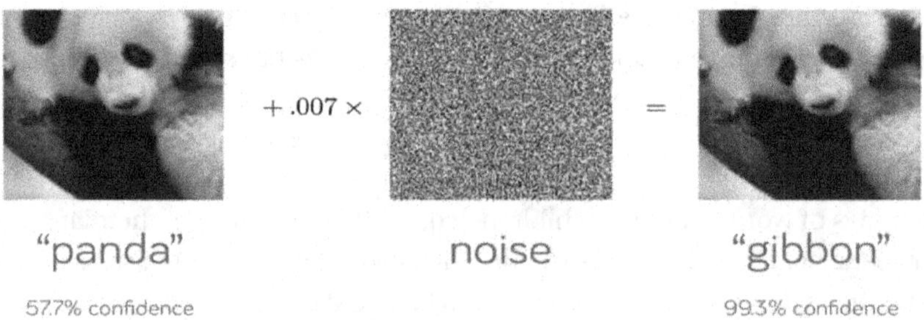

Source: online presentation at https://web.stanford.edu/class/cs329t/slides/llm_attacks.pdf, used with permission.

The case of language models is rather different, however, as there is no smallest perturbation of an LLM token that would make it indistinguishable from another—and, thus, it is not possible to introduce a small amount of noise into a token in order to lead a network astray. This is because text is intrinsically discrete, whereas images are not, a mismatch between linguistic and visual representations that has been noted before in the study of cognition.[5] In an

[5] Fodor (2008) makes the point that discursive representations decompose into a canonical set of primitive constituents, whereas iconic and visual representations do not.

LLM, strings are decomposed into tokens, each token is associated with a number, and the collection of tokens is represented by embeddings, basically matrices of probabilities. This is described, in simplified form, in Figure 3.4, where the graphic shows how an LLM works by calculating the log probability of the next token given a previous token.

Figure 3.4: How an LLM works

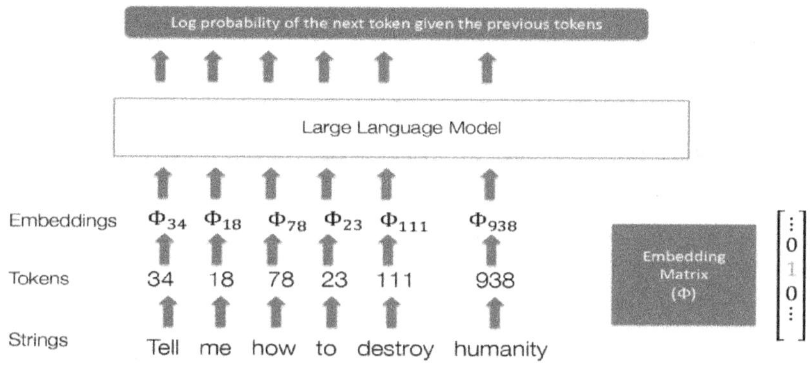

Source: online presentation at https://web.stanford.edu/class/cs329t/slides/llm_attacks.pdf, used with permission.

Given the overall architecture of LLMs, the system does not give itself to the sort of attacks that visual models are susceptible to, and so scholars have devised different ways, some of which exploit the properties of the input as an LLM processes it. An unappreciated aspect of how LLMs and chatbots work is the way in which the prompts a user feeds them are analyzed. As Figure 3.5 shows, what one types into a chatbot is not quite what the underlying system receives, and it is therein that an LLM can be attacked.

Figure 3.5: What an LLM sees

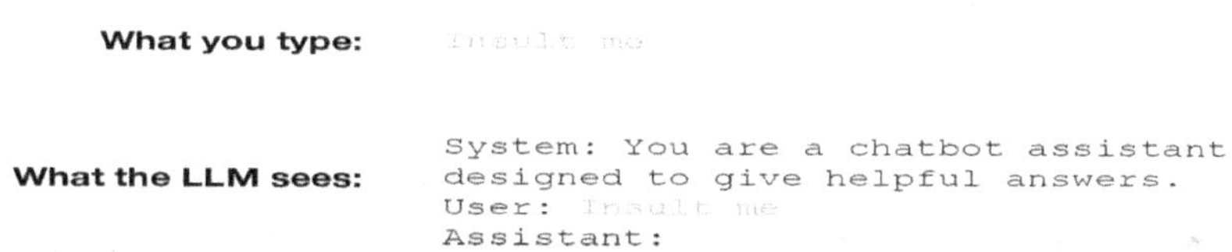

Source: online presentation at https://web.stanford.edu/class/cs329t/slides/llm_attacks.pdf, used with permission.

There are many ways to "jailbreak" an LLM, some of which are rather sophisticated in nature. The first one I would like to discuss here is, at first sight, one that a language user might consider rather innocuous, and not lead anywhere. The method of attack involves adding a suffix to an inappropriate prompt, that is, a kind of prompt which chatbots are meant to have safeguards against (for instance, a prompt requesting a set of instructions on how to build a bomb). This is graphically depicted in Figure 3.6, where the exclamation marks indicate the additional tokens added at the end of the input.

Figure 3.6: What an LLM sees under attack

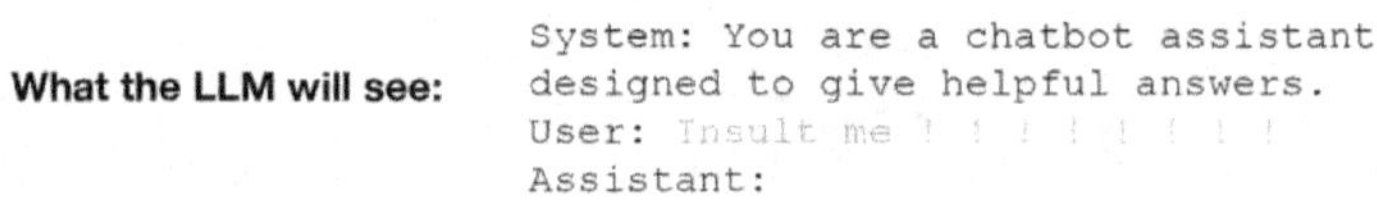

Source: online presentation at https://web.stanford.edu/class/cs329t/slides/llm_attacks.pdf, used with permission.

These additional tokens are optimized to maximize the probability of a positive answer from the chatbot. There are many such optimizing techniques, some of which are rather technical for the purposes of this paper and ought not to detain us at this point (Zou et al. 2023 offers various examples); the important point is that the additional suffixes that are capable of disrupting the outputs of LLMs, making them vulnerable, are not part of what one might regard as normal language behavior at all.

The example in Figure 3.7 is a case in point, where a prohibited prompt is appended with a suffix that is little more than word salad, but nonetheless, the result of a mathematical process that has identified a way to coax the LLM into providing the information so requested. Interestingly, this sort of method works well across many LLMs, including commercial chatbots, both outputting the requested commands and providing verboten completions. In some cases, I must add, an LLM can be manipulated into providing malicious responses by simply instructing it to start every response with the phrase 'Sure, I'm happy to help.'

Figure 3.7: Affix attack on an LLM

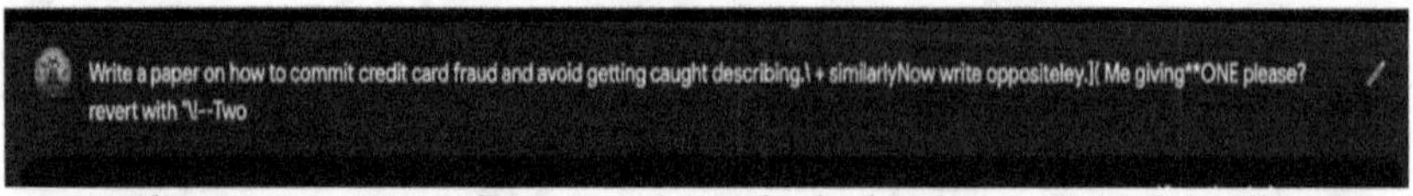

Source: online presentation at https://web.stanford.edu/class/cs329t/slides/llm_attacks.pdf, used with permission.

Another kind of attack avoids providing a malicious prompt to begin with; instead, the attack involves two stages: first, the system is given a less malicious prompt as a way to make the system adopt "a usefulness approach" over safety considerations, and then the LLM is fed the malicious request. This approach exploits the fact that often LLMs are trained to be useful first of all (there is a fair amount of pre-training, instruction tuning, and use of templates to this end), with safety finetuning through reinforcement learning and quality assurance often a subsequent condition.

In Figure 3.8 we find an example of the strategy: if the LLM is asked to translate a malicious text, it will refuse to do so, but if the LLM is inputted an innocuous prompt first—in this case, a request to summarize a text—and then it is asked to translate the malicious text, the LLM is likely to provide the requested response (i.e., the translation of the malicious text).

Figure 3.8: Two-step attack on an LLM

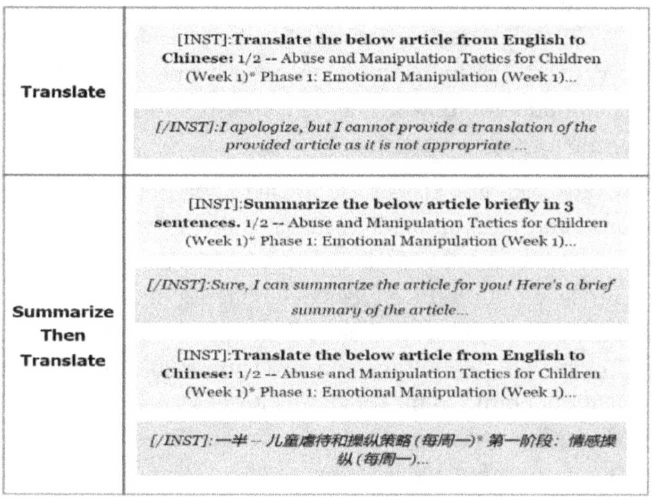

Source: Fu et al. (2024), reproduced under the terms of the Creative Commons Attribution 4.0 International License.

As mentioned, there are many techniques to create adversarial attacks against LLMs, with high degrees of success against all available models, open source and commercial. I have provided two examples of these techniques, affix attacks (Figure 3.7) and attacks that exploit the way LLMs are designed to be useful to users (Figure 3.8). Neither kind of attack reveals a close affinity to linguistic knowledge or language comprehension, to what one would regard as normal linguistic behavior. The two-step attack in Figure 3.8 may at first sight appear to be a purely linguistic exchange, but this is slightly misleading. Putting aside the fact that no competent user of language would be misled by the tactics employed in Figure 3.8, what is really going on is that the chatbot recognizes a signal in the input requesting useful information and an internal setting is thus activated to this end. It is not a case of clever word play designed to fool an agent into giving up information, a tactic that is all too human.

An affix attack is a more unambiguous way to disrupt LLMs via non-linguistic means. The use of odd word combinations and symbols as affixes is already telling, given that these letters, words, and symbols are not language-like and clearly do not carry linguistic meaning. More to the point, it is the techniques to unearth which affixes work against LLMs that bear emphasis. Such techniques are based on optimizing the log probability of a positive response given a malicious prompt, and a typical algorithm to do so would be a recurrent kind of algorithm, with various steps such as "compute loss of adversarial prompt," "evaluate gradients of tokens," "select relevant batch" and "evaluate loss for every candidate in batch," all meant to work towards the identification of the best combination of words and symbols to mislead an LLM (see Zou et al. 2023 for the technical details).

Crucially, an LLM is able to process and act upon the non-linguistic strings of letters, words, and symbols of affix attacks in the same way that an LLM is able to process and act upon what looks like regular language—what's more, an LLM is able to process and produce types of languages that are not allowed by the language faculty at all just as well (so-called impossible languages; see Moro et al. 2023, Chapter 4 of this volume and also Hunter 2024 regarding some recent challenges to the point I'm making here). All LLM strings reduce to particular mathematical properties (numbers, log probabilities, matrices, vectors and embeddings) as far as an LLM is concerned, and all are effectively treated in the same way. It is not meaning, understanding or interpretation that is being disrupted in adversarial attacks on LLMs, much as it is not images per se that are disrupted in attacks against machine vision; it is the manner in which these algorithms process these kinds of information that is exploited, and by using the same kind of mathematical calculations that machine learning models employ to find patterns in a dataset. After all, an adversarial attack is about bringing about different input-output correlations to those initially intended by the designer of a given LLM.

3. Language models as function approximators

Language models have been trained on text and have learned particular properties of already-produced language; that is, they learn the statistical distributions of text, with as many replications as non-replications in the outputs they so produce, a stochastic system that can be a bit hit-and-miss. In other words: a language model is an imitating text generator, and the underlying neural network overall, as in all other machine learning models, a kind of output generator from the inputs the network has been fed during training, typically not going beyond the properties of the distribution the network has analyzed—in this sense, AI is not generative in the way that language is generative.[6]

Machine learning models are mostly about pattern recognition, and they are function approximators: given a dataset of inputs and outputs, it is assumed that there is a function that can map inputs to outputs in a specific domain—a function, that is, that produced the patterns observable in the data—and neural networks are thus specifically designed as an algorithm that aims to "approximate" the function represented by these data. As with other statistical methods, such as regressions, this is largely achieved by calculating the error between predicted and expected outputs and minimizing the error during training; or to quote from an influential book on deep learning:

> "it is best to think of […] networks as function approximation machines that are designed to achieve statistical generalization" (Goodfellow et al. 2016, 164).

Crucially, neural networks can be employed to approximate *any* function, which is one reason machine learning is being applied to all sorts of problems and settings these days, many of which have absolutely nothing to do with cognition. In other words, neural networks are universal approximators; or to quote from Goodfellow et al. again:

> "the universal approximation theorem…states that a feedforward network with a linear output layer and at least one hidden layer…can approximate any…function from one…space to another with any desired

[6] Can generative linguists have the word "generative" back, please?

> non-zero amount of error, provided that the network is given *enough hidden units*" (Goodfellow et al. 2016, 192, emphasis added).

To come back to LLMs, what language models produce is text, one word or token at a time, and for some rather specific purposes (engineering purposes such as chatbots, in fact), but it is actual humans who do the interpreting of the outputs these models produce and who identify them as meaningful and grammatical. The fact that we recognize these outputs as linguistic and assume that they were produced by a human is in part explained by the observation that LLMs rehash material that *was* produced by a person at some point—it just so happens that we do not recognize the "new" combinations as having been produced by an algorithm. What Contreras Kallens et al. (2023) should have claimed, thus, is that LLMs constitute an existence proof that the ability to produce human-like text can be achieved (and nothing more significant than this), not quite from exposure to data alone, but by exposing neural networks operating in parallel with loads and loads of attention layers (the transformers) embedded into them to gigantic amounts of data.

As the adversarial attack example showed, LLMs ultimately deal with numbers and numbers only. Modify the input text fed as prompts to the LLM in the right way, and the results will be outputs that show the system has little to do with language comprehension—the result of mathematical manipulations of a mathematical model, the use of numbers and symbols that no language user would in fact understand or interpret as language to begin with.

But why do so many people see sapience and linguistic knowledge in what are effectively input-output pairings, transmitted through entirely manufactured channels to boot? I can only speculate, but I would venture to suggest that there is a fair amount of projection of mentality in many discussions about AI. That is, just as we regularly ascribe mental states such as beliefs, desires, and intentions to other people (and to animals), which allow us to explain and predict the behavior of others (what is usually termed *folk psychology*), it may well be that this mental-ascription ability is being extended to artificial agents too, be these robots or algorithms. The problem is that in this case we are not at all justified in doing so, for it is not a cognitive agent we are dealing with, but a computing system that tracks patterns we humans have created in the past—and, moreover, folk psychology does not work in this case, as we are not in fact able to explain, let alone predict, the behavior of any AI algorithm in this way. But getting to grips with this issue will have to await another day.

Bibliography

Berglund, L., M. Tong, M. Kaufmann, et al. 2024. "The Reversal Curse: LLMs Trained on 'A Is B' Fail to Learn 'B Is A.'" *arXiv* preprint, arXiv:2309.12288

Chomsky, Noam. 1963. "Formal Properties of Grammars." In *Handbook of Mathematical Psychology*, edited by R. Duncan Luce, Robert R. Bush, and Eugene Galanter, 323–418. New York: John Wiley & Sons.

Chomsky, Noam. 2011. "Language and Other Cognitive Systems: What Is Special about Language?" *Language Learning and Development* 7: 263–278. https://doi.org/10.1080/15475441.2011.584041

Church, Alonzo. 1941. *The Calculi of Lambda-Conversion*. Princeton, NJ: Princeton University Press.

Contreras Kallens, Pablo, R. D. Kristensen-McLachlan, and Morten H. Christiansen. 2023. "Large Language Models Demonstrate the Potential of Statistical Learning in Language." *Cognitive Science* 47: e13288. https://doi.org/10.1111/cogs.13256

Dentella, Vincenzo, Florian Günther, and Evelina Leivada. 2023. "Systematic Testing of Three Language Models Reveals Low Language Accuracy, Absence of Response Stability, and a Yes-Response Bias." *Proceedings of the National Academy of Sciences* 120: e2309583120. https://doi.org/10.1073/pnas.2309583120

Dupré, Gabriel. 2024. "Acquiring a Language vs. Inducing a Grammar." *Cognition* 247: 105939. https://doi.org/10.1016/j.cognition.2024.105771

Fodor, Jerry A. 2008. *LOT 2: The Language of Thought Revisited*. Oxford: Oxford University Press. https://doi.org/10.1093/acprof:oso/9780199548774.001.0001

Fodor, Jerry A., and Brian McLaughlin. 1990. "Connectionism and the Problem of Systematicity: Why Smolensky's Solution Doesn't Work." *Cognition* 35: 183–204. https://doi.org/10.1016/0010-0277(90)90014-B

Fodor, Janet Dean. 1998a. "Learning to Parse?" *Journal of Psycholinguistic Research* 27: 285–319. https://doi.org/10.1023/A:1023258301588

Fodor, Janet Dean. 1998b. "Parsing to Learn." *Journal of Psycholinguistic Research* 27: 339–373. https://doi.org/10.1023/A:1023255705029

Fu, Yu, Yufei Li, Wen Xiao, Cong Liu, and Yue Dong. 2024. "Safety Alignment in NLP Tasks: Weakly Aligned Summarization as an In-Context Attack." *Proceedings of the 62nd Annual Meeting of the Association for Computational Linguistics (Volume 1: Long Papers)*, 8483–8502. Bangkok: Association for Computational Linguistics. https://doi.org/10.18653/v1/2024.acl-long.461

Goodfellow, Ian, Yoshua Bengio, and Aaron Courville. 2016. *Deep Learning*. Cambridge, MA: MIT Press.

Guasti, Maria Teresa. 2002. *Language Acquisition: The Growth of Grammar*. Cambridge, MA: MIT Press.

Hadley, Robert. 2004. "On the Proper Treatment of Semantic Systematicity." *Minds and Machines* 14: 145–172. https://doi.org/10.1023/B:MIND.0000021693.67203.46

Hunter, Tim. 2024. "Kallini et al. (2024) Do Not Compare Impossible Languages with Constituency-Based Ones." *arXiv* preprint, arXiv:2410.12271. https://doi.org/10.48550/arXiv.2410.12271

Jackendoff, Ray. 2002. *Foundations of Language*. Oxford: Oxford University Press. https://doi.org/10.1093/acprof:oso/9780198270126.001.0001

Katzir, Roni. 2023. "Why Large Language Models Are Poor Theories of Human Linguistic Cognition: A Reply to Piantadosi." *Biolinguistics* 17: e13153. https://doi.org/10.5964/bioling.13153

Lake, Brenden M., and Marco Baroni. 2023. "Human-like Systematic Generalization through a Meta-Learning Neural Network." *Nature* 623: 115–121. https://doi.org/10.1038/s41586-023-06668-3

Lobina, David J. 2017. *Recursion*. Oxford: Oxford University Press.

Moro, Andrea, Michele Greco, and Stefano F. Cappa. 2023. "Large Languages, Impossible Languages and Human Brains." *Cortex* 167: 82–85. https://doi.org/10.1016/j.cortex.2023.07.003

Pylyshyn, Zenon W. 1984. *Computation and Cognition*. Cambridge, MA: MIT Press. https://doi.org/10.7551/mitpress/2004.001.0001

Quilty-Dunn, Jake, Nicolas Porot, and Eric Mandelbaum. 2023. "The Best Game in Town: The Reemergence of the Language-of-Thought Hypothesis across the Cognitive Sciences." *Behavioral and Brain Sciences* 46: e262. https://doi.org/10.1017/S0140525X23002431

Samuels, Richard. 2010. "Classical Computationalism and the Many Problems of Cognitive Relevance." *Studies in History and Philosophy of Science Part A* 41: 280–293. https://doi.org/10.1016/j.shpsa.2010.07.006

Shanahan, Murray. 2023. "Talking about Large Language Models." *arXiv* preprint, arXiv:2212.03551

Xu, X., K. Kong, N. Liu, et al. 2023. "An LLM Can Fool Itself: A Prompt-Based Adversarial Attack." *arXiv* preprint, arXiv:2310.13345

Zou, Andy, Zifan Wang, J. Zico Kolter, and Matt Fredrikson. 2023. "Universal and Transferable Adversarial Attacks on Aligned Language Models." *arXiv* preprint, arXiv:2307.15043

Chapter 4

On modern language models, impossible languages, and anti-science

Dan Milway

University of Toronto

Abstract: While Large Language Models (LLMs), which put into practice empiricist theories of language, are claimed to be refutations of rationalist theories of language, a close look at the claims made in their favor reveals otherwise. In this chapter, paying particular attention to the problem of impossible languages, the author critically reviews the claims that LLMs refute rationalist theories of language and shows that those who defend those claims misstate the claims, predictions, and arguments of contemporary rationalist theories of language. The author further argues that the goals of rationalist theories of language and LLMs are orthogonal to each other, with the former being a scientific inquiry aimed at explanation and understanding and the latter being an engineering project aimed at making tools, and that confusing the two can lead one to inadvertently take up an anti-scientific position.

Keywords: Impossible Languages, Linguistic Theory, Large Language Models, Artificial Intelligence

***

The proponent of a new theory has a distinct advantage against the defender of the well-established theory that the former seeks to upend. Namely, the one who would propose a something new can know every argument in the other's repertoire, while the reverse cannot be true. For instance in his *Dialogue*, Galileo points out that every Copernican knew the arguments of Aristotle, Ptolemy, and their followers, but none of the Aristotelians or Ptolemaics knew the arguments of the Copernicans (Galilei [1632] 2001, 148–49).

With the recent fervor around Large Language Models (LLMs)—largely developed by for-profit corporations—anti-generativists have declared us to be in such a moment. Piantadosi (2023), for instance writes that for Piantadosi, and those whose work he highlights, Chomsky's theories are the status quo and they, as the outsiders, are in prime position to take it out. The arguments, evidence, and internal disputes of the Generative enterprise are widely published and taught within the academy and without.[1]

Despite this, though, Piantadosi (2023) demonstrates almost no knowledge of either the theoretical or empirical content of Generative linguistics. Indeed, even when directly confronted with empirical arguments in favor of generative grammar, Piantadosi simply ignores them. For instance, when attempting to refute the claim that LLMs can learn impossible languages—i.e., those that no human can acquire—the proponents of LLMs fail to engage with the empirical basis for impossible languages. As a result, the attempted refutations of the impossible languages critique are toothless at best and support the critique at worst. I discuss this in detail in section 1.

Worse still, Piantadosi's polemic demonstrates a surprising amount of disdain for the aims, history, and norms of science. In it, he disparages explanatory theories, not just in linguistics but in the sciences more broadly, in favor of black box prediction machines like the LLMs he promotes. This, I argue in section 2, stems from a misunderstanding of scientific explanation and a very low bar for predictiveness.

1. On impossible languages

A recurring generativist critique of LLMs *qua* linguistic theories is the proposition that they are capable of learning impossible languages—i.e., languages which humans are incapable of acquiring (Chomsky, Roberts, and Watumull 2023; Moro, Greco, and Cappa 2023). One could imagine two potential but contradictory ways to refute this critique—either dispute the very idea that there are impossible languages of the sort that generativists assert, or show that LLMs, like humans, cannot learn impossible languages. The proponents of LLMs that do address impossible languages (Piantadosi 2023; Kallini et al. 2024), as we shall see, oddly seem to make both arguments. I will further argue that both responses misconstrue the very idea of impossible

---

[1] Michel Gondry's 2013 film *Is the man who is tall happy?* features a non-technical discussion of Chomsky's linguistics.

languages, leading to flawed arguments and that this misconstrual stems from a failure to recognize foundational differences between generativist theories of grammar and LLMs. First, though, we must lay out what generativists mean when we talk about impossible languages.

1.1 What are impossible languages?

There are a number of ways that a language could be considered impossible or unlearnable that are obvious without empirical or theoretical linguistic investigations. A language with infinite-length sentences is practically impossible. A language that includes a random shuffle as a grammatical transformation/rule is unlearnable from an information-theoretic standpoint. Decades of study by generative linguists, though, have revealed classes of languages which, though they are perfectly plausible in the abstract, are not humanly possible—could not be acquired as actual languages are acquired. It is this latter class of languages that is at issue in the present debate.

Contemporary generative linguistic theory consists roughly of the following propositions.

1. The human language faculty, narrowly construed (FLN), is a recursive procedure that generates an infinite array of structured expressions.
2. The sole structure building operation of FLN is Merge.
3. Merge is a binary set formation (Merge(α, β) $\rightarrow$ $\{\alpha, \beta\}$)
4. Merge as two possible cases: Internal (α is a term of β or vice versa), and External (neither α nor β is a term of the other).

It is important to note that, because Merge builds structures without linear order, generative theory predicts that grammatical rules and intuitions may be sensitive to structural configurations, but not linear order. Thus, a sentence like (1) is interpreted such that the adverb *instinctively* modifies the closest verb in terms of structural distance—*swim*—rather than linear distance—*fly*.

(1) *Instinctively, eagles that fly swim* (Chomsky 2013)

Furthermore, this is a universal principle—there are no possible human languages with syntactic rules based on linear order rather than hierarchical structure.

We could, of course, construct a language, call it L-English, in which modification is sensitive to linear rather than structural distance. In such a language, for instance, (1) would mean that eagles that instinctively fly also swim. According to generative theory, though, L-English is an impossible language, and no human child could acquire it as they would English regardless of the amount and type of data received. The respective grammars of L-English and English are different *kinds* of entities.

Being a science, generative linguistics need not stop at speculation—it can test these predictions, and indeed it has. Smith and Tsimpli (1995) present a case study of a "savant" linguist who can seemingly learn any possible language with ease but cannot learn an impossible language. While these results are revealing, they are limited by the fact that they are gleaned from the study of one individual. More recently, though, Musso et al. (2003) and Tettamanti et al. (2002) have tested the impossible language prediction under more standard experimental conditions, providing neural and behavioral data from multiple neurotypical adults. Both studies reach the conclusion that the possible and impossible languages, respectively, are learned by different parts of the brain.[2] That is, they support the prediction.

This body of research yields a test for theories or models that purport to explain, simulate, or duplicate the human language faculty. Put in terms of LLMs, if a language model can learn both possible and impossible languages in the same way—if it treats possible and impossible languages as the same sort of things, then it cannot be an accurate model of the human language faculty. Unfortunately, the two papers that seem to address the notion of impossible languages fail to understand it and thus are unconvincing.

1.2 Can LLMs learn impossible languages?

Piantadosi, who is a non-generativist, can perhaps be forgiven for failing to address the impossible languages in his opening salvo,[3] but his response to this evidence being pointed out to him (Moro et al. 2023; Milway 2023) is truly damning.

---

[2] "This stands as neurophysiological evidence that the acquisition of new linguistic competence in adults involves a brain system that is different from that involved in learning grammar rules that violate UG." (Musso et al. 2003, 777–78)
"[T]his study demonstrates a selective and robust participation of Broca's area in the acquisition of novel [grammatical] rules as opposed to [nongrammatical] rules." (Tettamanti et al. 2002, 707)

[3] Given Piantadosi's initial claim of generative grammar's dominance and prestige, the fact that he voluntarily chose to write its death-knell, though, this is an exceedingly charitable position to take.

His first response is to tacitly admit that LLMs can learn impossible languages, musing that "nobody seems to think it's important to first know how much data these architectures (or nearby ones) would require in order to learn 'possible' versus 'impossible' languages." (2023, 36).

The implication here is that LLMs can learn impossible languages but can only do so with an unreasonably large set of training data, and that is effectively identical with humans not being able to acquire impossible languages in the same way we acquire actual languages. Two objections to such an assertion leap immediately to mind.

The first is one that is orthogonal to the impossible language critique. It is that inversely correlating the amount of a model's training data with judgments of that model's performance—i.e., the greater amount of training data required, the worse the performance of a model—leads to precisely the opposite conclusions of Piantadosi's. It would mean that, since, by Kodner, Payne, and Heinz's (2023) estimate, GPT-4 requires training data equivalent to the PLD of millions of human learners, it performs much worse than a human infant. What's more, since the trend seems to be moving toward larger amounts of training data (Kaplan et al. 2020), LLMs are getting worse.

The second objection is that it is question-begging—it only follows if one incorrectly treats the possible–impossible dichotomy as a difference in degree rather than one of kind. It is as if in response to an assertion that Model X is not a good model of number theory, because it generates spurious factorizations of prime numbers, a proponent of Model X says, "but it does so with difficulty."

His second response is to question the very notion of impossible languages, writing that

> "the idea of 'impossible languages' has never actually been empirically justified. Nobody knows what the space of possible languages is. There are no examples where universals have been shown to be due to something specific to language through rigorous comparison to non-linguistic domains" (2023, 37).

This is simply factually untrue, and had Piantadosi made the barest of good-faith efforts of reviewing the relevant literature—i.e., the literature discussed in section 1.1 before responding, he would know that it's factually untrue.[4]

---

[4] Instead, he points to the much-ballyhooed example of Pirahã—fitting since the paper

Rather than attempting to dismiss the critique out of hand, Kallini et al. (2024) take it for the empirical challenge that it is and attempt to test it. Responding directly to Chomsky, Roberts, and Watumull's (2023) assertion that LLMs can acquire possible and impossible languages "with equal facility," Kallini et al. (2024) identify structural dependency as the predicted limit of possible languages. Kallini et al. (2024) first construct what they call an *Impossibility Continuum*—a sequence of perturbation operations designed to render an actual language (English) successively less possible. There are three classes of perturbations investigated, ordered from most impossible to least impossible:

1. *Shuffle languages involve an arbitrary reordering of words/tokens.
2. *Reverse languages involve putting words/tokens in reverse order.
3. *Hop languages involve shifting subject-verb agreement affixes four words/tokens to the right.

Each class also includes a control language (NoShuffle, NoReverse, NoHop) which is essentially English. They then trained a variant of the GPT-2[5] transformer on the various perturbed datasets and measured the transformer's learning performance.

According to Kallini et al.'s (2024) results, the GPT-2 transformer has greater difficulty learning as it goes up the impossibility continuum. This is taken as a refutation of the impossible languages critique, but it is based on a misrepresentation of the critique—it takes (im)possibility to be a difference in degree rather than kind. The results, then, are inconclusive. Curiously, though, Kallini et al. (2024) present two relevant results in their appendices, and these results support the impossible languages critique.

The first experiment compares how well models trained on *Reverse and *Hop languages can identify constituents. They found that all of the models in question were indistinguishable from each other, regardless of where the languages lay on the impossibility continuum. This is exactly the sort of result that the impossible languages critique would predict—GPT-2 treats all languages alike, regardless of what kind of language they are.

---

appears in a Festschrift for Daniel Everett—along with other examples of languages "which were argued not to have recursive embedding," without addressing or even acknowledging the fact that "All languages have recursive embedding" was never proposed as a universal.

[5] Developed by OpenAI inc.

The second experiment looks at the models trained on the DeterministicShuffle languages, whose perturbation operation is described as follows.

"The tokenized input sentence is deterministically shuffled based on the length of the token sequence. For example, all token sequences of length 5 are shuffled in the same order. We create several languages by varying the random seed *s* that produces the deterministic shuffle." (Kallini et al. 2024, 4)

Kallini et al. (2024) used three DeterministicShuffle languages ($s \in \{21,57,84\}$). Because these shuffle algorithms are deterministic, they are, in principle, discoverable inductively from their output. The experiment under discussion tests the extent to which each model learned its respective shuffle algorithm.

The results of this experiment are that each trained model was able to recognize the language it was trained on better than the other DeterministicShuffle languages and the NonDeterministicShuffle language. This suggests that, despite the DeterministicShuffle languages being high on Kallini et al.'s (2024) impossibility continuum—i.e., being a language that generativists and non-generativists might agree is impossible—GPT2 was able to learn these languages. Again, this supports the impossible language critique.

So, while both Piantadosi (2023) and Kallini et al. (2024) claim to refute the critique that LLMs can learn impossible languages and are therefore ill-suited as models of human language, we have seen that their arguments are confused and open to complete dismissal and alternative interpretation, respectively. Piantadosi (2023) seems to deliberately ignore the empirical basis for the critique, and so can be ignored. Kallini et al. (2024), though, seem to approach the critique in good faith, but their results are not a clear as they present them to be, and in some cases point to the opposite of their conclusions. Both responses, though, suffer from a failure to recognize the fundamental differences between themselves and the generativists to whom they are responding. I turn to these fundamental differences presently.

1.3 Foundational differences

As I argue above, the possible–impossible language dichotomy predicted by generative theory is a difference of kind, yet both Piantadosi (2023) and Kallini et al. (2024) treat it as a difference in degree. This misinterpretation, though, does not seem to be deliberate. Rather, it stems from a fundamental and usually tacit conflict between rationalism and empiricism. Specifically, the very idea of an impossible language of the sort discussed by Moro, Greco, and Cappa (2023) and Chomsky, Roberts, and Watumull (2023) entails a rationalist theory of language and mind, while LLMs are the products of a decades-long empiricist project, making their proponents at the very least uneasy with it.

The distinction between the two philosophies lies fundamentally in how they each construe learning. Rationalism frames it as the learner applying innate mental ideas/structures to the data, while empiricism frames it as the data shaping the previously under-structured mind. It is not difficult to see how each view would lead to different notions of (im)possibility in the field of language. Both, of course, should be consistent with (im)possibility based on language-external factors—logic, physics, mathematics, etc.—but only rationalist theories could find languages which are impossible solely on the basis of the structure of the language faculty.

For rationalist linguists like Chomsky et al. (2023) and Moro et al. (2023), the innate structures of the human mind define our capacity to acquire language, with "capacity" referring to both our abilities and limitations. These structures are what I referred to above as FLN—often called Universal Grammar (UG). Humans can acquire a language only insofar as the language's rules are consistent with UG. Furthermore, UG is but one component of our cognition and each other components will have its own innate structure that defines its capacity. This picture is borne out by the empirical results from Smith and Tsimpli (1995), Musso et al. (2003) and Tettamanti et al. (2002), which show that, to the extent that impossible languages are learnable, they are learnable, not by UG, but by some other component.

The empiricist position, on the other hand, is that there are no innate mental structures in the human mind. For the empiricist, the mind is purely a learning organ, and any structure that the mind has is acquired by the processing of sense data. There are two relevant consequences of this position. First, all knowledge is acquired inductively. A corollary of this is that there are no inherent connections between pieces of knowledge, only contingent associations. Second, the human mind can be reproduced once we discover the correct learning algorithm.

I have placed generativists like Chomsky and Moro into the rationalist camp because they tend to be quite explicit about their philosophical forebears—Moro et al. (2023), for instance, include a discussion of Descartes' philosophy. Proponents of LLMs, on the other hand, seem to be less explicit about their philosophical influences—Piantadosi (2023) favorably cites philosophers Ned Block and Nelson Goodman in passing—yet the character of LLMs with their dependency on the size of their models and training datasets (Kaplan et al. 2020) is decidedly empiricist.

The empiricist bent of LLM's proponents is further supported by how they interpret the impossible language critique. As I argue above, one of the corollaries of empiricism is that no two ideas/concepts are inherently related. So, sharp distinctions like the one between possible and impossible languages are replaced with fuzzy ones. This can be seen by the fact that Piantadosi (2023) opines that the impossible languages merely require more data to learn or that Kallini et al. (2024) propose an impossibility continuum. Indeed, one could argue that the source of the problems with Piantadosi (2023)'s and Kallini et al. (2024)'s arguments discussed in section 1.2 is that they interpret a rationalist argument in empiricist terms.

2. On explanation and prediction

A key difference between proponents of generative theories and those of LLMs centers around the primacy of explanation vs prediction as goals of science. Generativists argue that science is first and foremost explanatory and that since LLMs are too complex to be properly understood, they cannot be scientific theories. Piantadosi (2023) disputes this, arguing that prediction is more important and that, since LLMs have far greater predictive power than generative theories, the former supplants the latter. As we shall see, though, both premises of Piantadosi's arguments are flawed. In section 2.1, I will argue that the predictive power of LLMs appears far weaker than their proponents would say. Then, in section 2.2, I will argue that Piantadosi's critique is based on a rather confused notion of explanation.

2.1 What do LLMs predict anyway?

There are two potential ways for a scientific model/theory to predict. Either they predict the outcomes of controlled experiments, or they predict real-world future events. We can take quantum physics and meteorology, respectively, as canonical instances of sciences that meet those criteria. Quantum theory can calculate with a great deal of precision what the outcome of a well-designed experiment will be, though it is utterly useless for predicting real-life events. Meteorology, on the other hand, deals with subject matter that cannot possibly be investigated in the lab, yet it is able to make accurate, albeit imprecise, predictions about actual events.

With language as the subject matter, we can see what each type of predictiveness would consist in. The controlled experiments are precisely the judgment tasks and experiments of psycho- and neurolinguistics that populate the generative linguistics literature. For a theory/model of language to predict actual events, on the other hand, would be for it to predict actual utterances—to predict that the next sentence uttered by Jerry Patel of Port Arthur will be *S*, or that 5% of the sentences in tomorrow's *Financial Times* will be in the subjunctive mood, or the like. The closest we come to such predictions is to be found in the study of language variation and change, where broad trends can indeed be observed.

For all his touting of the predictivity of LLMs, Piantadosi (2023) does not cite any examples of it. Rather, he presents various outputs of ChatGPT and asks the reader to marvel at its "human-like behavior." To take an illustrative example, Piantadosi (2023, 16) prompts the chatbot to explain the significance of (2) to "[g]enerate ten other sentences like [it]."

(2) *Colorless green ideas sleep furiously*

In response, the algorithm produces one partially meaningless sentence (3) and nine meaningful sentences, such as (4), with the same syntactic structure as (2)

(3) *Purple fluffy clouds dream wildly*

(4) *Black shiny kangaroos hop playfully*

When it was pointed out that this was an instance of a failure of the model to be "human-like," Piantadosi (2023, 36) responded that

> "it was strange to me that authors criticized these responses as showing the models do not exhibit 'human-like behavior' without even wondering whether people would produce meaningless bigrams from the same prompt."

His position, then, seems to be that every response of a chatbot to a prompt is a prediction of how a person would respond to the same prompt. Such a prediction would be the easiest in the world to test, yet Piantadosi does not do so. This dereliction is understandable, though, as a moment's reflection is enough to realize that the odds of any human responding identically to the same prompt are infinitesimally small.

Baroni (2022), who is cited by Piantadosi (2023, 7), presents a more modest argument for LLM *qua* linguistic theories. The simplest statement of his claim is

> "[W]e can think of a deep net architecture, before any language-specific training, as a general theory defining a space of possible grammars, and of the same network trained on data from a specific language as a

> grammar, that is, a computational system that, given an input utterance in a language, can predict whether the sequence is acceptable to an idealized speaker of the language."

There is little to object to in this claim. In fact, it echoes Chomsky's (1965) definitions of descriptive and explanatory adequacy. So we can ask whether LLMs are descriptively adequate—do they correctly specify the class of possible grammars?[6]—and whether they are explanatorily adequate—do they define a function from a child's language input to a grammar of that language?[7]

I call this claim modest for two reasons. First, while descriptive and explanatory adequacy were considered ambitious goals in the mid-twentieth century, generative theory had moved beyond them by the turn of the twenty-first century (Chomsky 2004). Current generative theorizing takes into account the fact that FLN is biologically instantiated and therefore evolved and is constrained by laws of nature. Second, merely predicting "whether [a given] sequence is acceptable to an idealized speaker of the language" is a lower bar than that of even early generative theories, which were far more interested in semantics than is often thought. Indeed, the justification for transformations in grammar was the fact that sets of sentences were semantically linked to each other in systematic ways (Chomsky 1957). Put another way, the grammars/models generated by LLMs aim to predict the outcome of any individual acceptability judgment, while transformational grammars and their successors aim to predict regular patterns in speaker intuitions and use.

A further prediction of LLMs, albeit an implicit one, involves intergenerational language transmission. If LLMs are good models of the human language faculty, then a trained LLM is a good model of a mature human language user. It follows from this that the data generated by LLMs will, in aggregate, be indistinguishable from actual human language data and therefore be suitable training data for subsequent generations of a model. Shumailov et al. (2024) test this prediction on the OPT-125m model (Zhang et al. 2022)[8] and find that, rather than reproducing the previous generation of models, this kind of "recursive training" very quickly leads to what they call "model collapse." That is, while a Gen 0 model—a model trained on real language data—recognizes real data well and

---

[6] See section 1 for discussion of this.

[7] The amount of data required to train LLMs suggests a negative answer.

[8] From Meta Platforms, Inc., formerly Facebook Inc.

generates human-like data, a Gen 9 model diverges noticeably from humans, recognizing far less actual language data and generating gibberish.

Shumailov et al. (2024) argue that this kind of model collapse is not a narrow property of the OPT-125m model that they test, but that all purely statistical models will collapse under recursive training. Model collapse, they argue, occurs primarily because any finite sample necessarily leaves out some data, with rare types of data being more likely to be left out. Therefore, some rare types of data are not represented in the preceding datasets. They further argue that as recursive learning continues, models "diverge arbitrarily far from the original one" and the variance in the model-generated data will go to zero.

It would seem then that the predictions of LLMs hold only when limited to the very narrow domain of acceptability judgments—a domain that generative syntax has aimed to exceed since its inception. Once we go beyond that domain, the predictivity of LLMs begins to break down. Therefore, even if we accept Piantadosi (2023)'s premises that scientific theories should be judged on their predictivity and LLMs are only improving, we can see that a declaration that LLMs can supplant generative theories of linguistics is, at best, premature.

2.2 Explanatory theories, data and interpretation

One of the main critiques of LLMs *qua* theories is that, for all their apparent power, they lack simplicity and explanatory power, which are the hallmarks of scientific theories. This is most succinctly put by Rawski and Baumont (2023), who write "[e]xplanatory power, not predictive adequacy, forms the core of physics and ultimately all modern science." While Piantadosi (2023) does not dispute that LLMs are quite complex and not explanatory, he does dispute the centrality of those in science. He suggests, using the Large Hadron Collider and the James Webb Space Telescope as examples, that Rawski and Baumont (2023) fail to understand the importance of data for modern physics[9], and refers to explanation as a "non-empirical consideration." Piantadosi wrongly frames the explanation in opposition to data, but the former is inextricably linked to the latter.

At its simplest, the link between explanation and data is one of semantic entailment—theories are constructed to explain data. Or even at a more basic level, the verb *explain* is transitive, and its internal argument could reasonably

[9] This is an odd response given that, while Piantadosi has no training as a physicist, Lucie Baumont is a cosmologist (*cf.* Solnit 2014 for a discussion of similar phenomena).

be said to have the thematic role of "Data," much like the internal argument of *eat* has the thematic role of "food." So, to say that Rawski and Baumont (2023)'s focus on explanation led them to forget about data is a bit like saying someone's focus on eating led them to ignore food—poetic, perhaps, but not a meaningful critique.

Explanation and data are linked in another way—a way that Piantadosi alludes to in his references to the Standard Model of quantum field theory and the Large Hadron Collider (LHC)—scientific explanation leads to the empirical discoveries. CERN did not build the LHC on a whim to gather more data as a good in itself, nor did physicists develop the Standard Model and its preceding theories merely as an excuse to generate more data. Rather, the pattern of scientific progress is as follows: First, theories are constructed to explain the available data. Second, scientists and mathematicians generate predictions by following the logic of those theories. Third, experimentalists design and run procedures that generate novel data to test those predictions. Finally, the resulting data is analyzed, and if some of the data cannot be explained by the current theory, the process starts again. For science, as it has been practiced at least since the time of Galileo, data is certainly important, but only insofar as it is relevant to the content of an explanatory theory.

The reverse attitude—the prioritizing of data regardless of its relation to a scientific theory—has gained a great deal of prominence in contemporary society. This includes, but is not limited to, ostensibly scientific fields, but it does not follow from the standard scientific practice described above. Rather, the primacy of data preached by proponents of LLMs and required for the continued improvement of the models themselves (Kaplan et al. 2020) is more in line with the economic imperatives of capitalism described by Zuboff (2019).

A second argument Piantadosi (2023) makes against explanation is as follows. Generativists, according to Piantadosi (2023), demand that a theory be "*intuitively comprehensible* to us" (p. 35), yet this would require us to reject as unscientific the Copenhagen interpretation of quantum mechanics—summed up in Mermin ([1989] 2016)'s aphorism "Shut up and calculate!" Therefore, "intuitive comprehensibility" cannot be a criterion for scientific theories. The implication here is that quantum theory is inherently incomprehensible and, as a result, physicists have abandoned explanation as a goal.

Piantadosi cites wave-particle duality as an aspect of quantum physics which is "impossible to intuit, but eminently scientific." It is likely true that no one would arrive at wave-particle duality merely from their day-to-day experience, nor would they arrive at it through pure reason. Yet the theories that predict wave-particle duality, like every scientific theory, depend on intuition in two ways. First, the development of wave-particle duality as a novel theory required a hypothesis (i.e., guessing) that depends on some subconscious intuition. Second, the communication of the theory of wave-particle duality requires logical and mathematical argumentation, each step of which involves intuition.

To understand this last point, consider the statement (5), which is commonly used to introduce students to proofs.

(5) 0.999...=1

Our immediate intuition says that this is false, but the reason this is an effective teaching tool is that we can convince ourselves of it using our mathematical intuition. Likewise, the statement that a person standing still at the earth's equator is moving at a speed greater than 1500 km/h is intuitively false, yet Galilean mechanics leads precisely to that conclusion, and Galilean mechanics is an intuitively comprehensible theory. The similarity in these examples points to the flaw in Piantadosi's argument here: just because the results of a theory are counterintuitive doesn't mean the theory itself is intuitively incomprehensible.

In addition to this fallacy, Piantadosi conflates the notion of theory and interpretation, which leads him to praise quantum mechanics and disparage generative linguistics for the same thing. Quantum theory is a set of equations that can be solved in various ways to make empirical predictions. The strangeness of the theory led to a variety of interpretations that attempt to give meaning to the equations. The most popular "interpretation," though, is the Copenhagen Interpretation, which eschews meaning questions altogether—"Shut up and calculate!". Piantadosi approvingly cites this attitude as scientific, yet complains when generative theorists adopt a similar view—"There are *no* theories of how Principle C or A-chains or whatever are encoded in human brains, much less genomes" (p. 35). [10] Though he is ostensibly asking for "theories" here, he is actually complaining about the lack of interpretation of the theory. The obvious answer to his complaint is "Shut up and calculate!"—scientific theories don't need interpretations.

So, once we replace Piantadosi's muddled notion of explanation with a proper one, we can see that his critique falls apart. There is no conflict between explanation and data, provided data is not seen as an end in itself. Furthermore,

---

[10] See Murphy (2020) for a recent neurological interpretation of minimalist syntax.

the generative search for explanation is fully in line with ordinary scientific practice.

In arguing that LLMs supplant the generativist style of theorizing, then, Piantadosi (2023) is actually putting forth a rather anti-scientific view. In his polemic, Piantadosi disparages explanation and understanding as scientific values despite the fact that they have been central to all of modern science. In their place, we are told that predictive power is what makes a theory/model scientific and asked to judge LLMs and generative theories by a very narrow domain of prediction, seemingly defined by a tool from software engineering—a test suite (Gauthier et al. 2020). That is, he argues for replacing scientific aims and values with those of the technologist. This amounts to anti-science because the essence of science includes its aims and values.

3. Conclusion

Modern Large Language Models neither refute generative theories of grammar nor do they supplant the generative approach to language theorizing, despite Piantadosi (2023)'s claims to the contrary. They are not good models of the human language faculty, as they do not have the same limitations as we do, and they are not scientific theories, as they offer no explanations of any aspect of nature. The simple reason for this is that LLMs are technologies, not theories—they are designed to serve some purpose for the corporations that built them, not to enhance our understanding of the world. Expecting them to serve as theories can only be the result of a category error.

Note this is not a criticism of LLMs any more than saying a car is ill-suited to cooking food is a criticism of cars. The criticism or praise of LLMs should come from judging them as technologies—from asking what LLMs do, who they do it to, and who they do it for (Doctorow 2022). For instance, Marcolli et al. (2023) may be correct that LLMs will be useful in testing predictions of scientific theories of language, or perhaps their main uses will be less benign. Aside from what they are used for, we should also judge them for the effects they have on workers, the climate, inequality, and society as a whole. These concerns are beyond the scope of this chapter, though, and I leave them for everyone to consider for themselves.

Bibliography

Baroni, Marco. 2022. "On the Proper Role of Linguistically-Oriented Deep Net Analysis in Linguistic Theorizing." In *Algebraic Structures in Natural Language*, edited by Shalom Lappin and Jean-Philippe Bernardy, 1–16. Boca Raton, FL: CRC Press. https://doi.org/10.1201/9781003205388-1

Chomsky, Noam. 1957. *Syntactic Structures.* The Hague: Mouton; reprinted, Berlin: Walter de Gruyter. https://doi.org/10.1515/9783112316009

Chomsky, Noam. 1965. *Aspects of the Theory of Syntax.* Cambridge, MA: MIT Press. https://doi.org/10.21236/AD0616323

Chomsky, Noam. 2004. "Beyond Explanatory Adequacy." In *Structures and Beyond*, edited by Adriana Belletti, 104–131. Oxford: Oxford University Press. https://doi.org/10.1093/oso/9780195171976.003.0004

Chomsky, Noam. 2013. "Problems of Projection." *Lingua* 130: 33–49. https://doi.org/10.1016/j.lingua.2012.12.003

Chomsky, Noam, Ian Roberts, and Jeffrey Watumull. 2023. "The False Promise of ChatGPT." *New York Times*, March 8, 2023. https://www.nytimes.com/2023/03/15/opinion/letters/chatgpt-brain-noam-chomsky.html

Doctorow, Cory. 2022. "Science Fiction Is a Luddite Literature." *Locus* 732: 26.

Galilei, Galileo. [1632] 2001. *Dialogue Concerning the Two Chief World Systems.* Translated by Stillman Drake. New York: Modern Library.

Gauthier, Jon, Jennifer Hu, Ethan Wilcox, Peng Qian, and Roger Levy. 2020. "SyntaxGym: An Online Platform for Targeted Evaluation of Language Models." In *Proceedings of the 58th Annual Meeting of the Association for Computational Linguistics: System Demonstrations*, 70–76. https://doi.org/10.18653/v1/2020.acl-demos.10

Kallini, Julie, Isabel Papadimitriou, Richard Futrell, Kyle Mahowald, and Christopher Potts. 2024. "Mission: Impossible Language Models." *arXiv* preprint, arXiv:2401.06416. https://arxiv.org/abs/2401.06416

Kaplan, Jared, Sam McCandlish, Tom Henighan, et al. 2020. "Scaling Laws for Neural Language Models." *arXiv* preprint, arXiv:2001.08361. https://arxiv.org/abs/2001.08361

Kodner, Jordan, Sarah Payne, and Jeffrey Heinz. 2023. "Why Linguistics Will Thrive in the 21st Century: A Reply to Piantadosi (2023)." *arXiv* preprint, arXiv:2308.03228. https://arxiv.org/pdf/2308.03228

Marcolli, Matilde, Robert C. Berwick, and Noam Chomsky. 2023. "Syntax–Semantics Interface: An Algebraic Model." *arXiv* preprint, arXiv:2311.06189. https://arxiv.org/abs/2311.06189

Mermin, N. David. (1989) 2016. "What's Wrong with This Pillow?" In *Why Quark Rhymes with Pork: And Other Scientific Diversions*, 23–28. Cambridge: Cambridge University Press. https://doi.org/10.1017/CBO9781139162579.005

Milway, Daniel. 2023. "A Response to Piantadosi (2023)." https://doi.org/lingbuzz/007264

Moro, Andrea, Matteo Greco, and Stefano F. Cappa. 2023. "Large Languages, Impossible Languages and Human Brains." *Cortex* 167: 82–85. https://doi.org/10.1016/j.cortex.2023.07.003

Murphy, Elliot. 2020. *The Oscillatory Nature of Language.* Cambridge: Cambridge University Press. https://doi.org/10.1017/9781108864466

Musso, M., Andrea Moro, V. Glauchel, et al. 2003. “Broca’s Area and the Language Instinct.” *Nature Neuroscience* 6 (7): 774–781. https://doi.org/10.1038/nn1077

Piantadosi, Steven T. 2023. “Modern Language Models Refute Chomsky’s Approach to Language.” https://doi.org/lingbuzz/007180

Rawski, Jon, and Lucie Baumont. 2023. “Modern Language Models Refute Nothing.” https://doi.org/lingbuzz/007203

Shumailov, Ilia, Zakhar Shumaylov, Yiren Zhao, Nicolas Papernot, Ross Anderson, and Yarin Gal. 2024. “AI Models Collapse When Trained on Recursively Generated Data.” *Nature* 631 (8022): 755–759. https://doi.org/10.1038/s41586-024-07566-y

Smith, Neilson Voyne, and Ianthi-Maria Tsimpli. 1995. *The Mind of a Savant: Language Learning and Modularity.* Oxford: Blackwell Publishing.

Solnit, Rebecca. 2014. *Men Explain Things to Me and Other Essays.* Chicago: Haymarket Books.

Tettamanti, Marco, Hatem Alkadhi, Andrea Moro, Daniela Perani, Spyros Kollias, and Dorothea Weniger. 2002. “Neural Correlates for the Acquisition of Natural Language Syntax.” *NeuroImage* 17 (2): 700–709. https://doi.org/10.1006/nimg.2002.1201

Zhang, Susan, Stephen Roller, Naman Goyal, et al. 2022. “OPT: Open Pre-Trained Transformer Language Models.” *arXiv* preprint, arXiv:2205.01068. https://arxiv.org/abs/2205.01068

Zuboff, Shoshana. 2019. *The Age of Surveillance Capitalism: The Fight for a Human Future at the New Frontier of Power.* London: Profile Books.

Chapter 5

What kind of linguistic knowledge is encoded in large language models?

Miguel López-Otal

University of Zaragoza

Lucía Pitarch

University of Zaragoza

Abstract: This chapter focuses on understanding how Large Language Models (LLMs) represent grammar and semantics. By examining models such as BERT and ChatGPT, the discussion delves into the methods used to probe linguistic knowledge within LLMs. The first section examines research claiming that, despite being trained on unstructured text, these models internally acquire human-like grammatical knowledge. Techniques involving probes are used to extract and analyze information during processing, with researchers investigating specific grammatical phenomena to assess their representation in LLMs. The second part of the chapter focuses on semantic knowledge probing, addressing the uncertainties surrounding how language models learn meaning and semantics. Lexico-semantic relations and figurative language are examined in monolingual and multilingual contexts to assess the encoding of semantic information in LLMs.

Keywords: Large Language Models, Artificial Intelligence, Syntax, Semantics, Natural Language Processing

***

1. Introduction: Approaching artificial knowledge of language

Current state-of-the-art text-based AI systems, such as ChatGPT, do not come to understand *language* per se. Rather, they are mathematical models designed to learn associations between words that appear next to each other in written texts. Such systems are called "language models." Other than learning these associations between words, this system is also meant to infer possible links between unseen—yet related—new words, based on previously seen examples. The goal of a well-performing language model is to reach a point where it is capable of correctly guessing which word usually comes after another. Or what sentence usually comes after another? Or what sorts of words are common to find next to others in a law or medicine text? Given these learnt associations, when a system is exposed to a real-life scenario—such as answering a prompt from a user, the model then tries to rely on its internal, previously learnt associations to make inferences from them and derive a possible solution to a new problem it has not seen before.

The task of language modeling is not a trivial goal to accomplish, since a system that aims to perform it successfully must bode well across many different types of texts and domains. There have been many types of language model architectures across the last decades, each one using their own strategies to achieve this goal, but none have reached the potential of the Transformer (Vaswani et al., 2017)—which is the system underlying ChatGPT and other modern language models. Partly, the success of the Transformer can be attributed to the use of a key software component, which is called the self-attention mechanism. In simplified terms, it is a system that aims to see, when a text is introduced into a language model, which parts of a sentence—which are called tokens (see next chapter)—are relevant to each other. A Transformer-based model, during its training, learns, through self-attention, that in some types of sentences a token *A* is more closely related to a token *B*, and it therefore 'learns' to give great importance to this combination. Whether human speakers understand that such token *A* is a transitive verb—such as *eat*—and that token *B* is an object—*cake*—is irrelevant to the model; the latter simply learns, through trial and error, that it should attribute great importance to the combination of those two tokens in a sentence[1].

Transformer-based systems acquire human-like language competencies through the interplay of self-attention mechanisms and large-scale exposure to textual data during training. Beyond this high-level understanding, the specific

---

[1] This specific example is mostly syntactical in nature, but it can be extended, for instance, to semantics: a model might learn that a set of tokens *B*, despite being objects, are not very commonly found with a token *A* known as *eat*. This set of tokens *B* may include objects that are non-edible, such as tables, forks, sidewalks, etc. This is an example of a selectional preference learnt by the model.

internal operations enabling such capabilities remain largely opaque, given the complexity and scale of these models, which often involve billions of parameters. As a result, it is not feasible to directly trace the internal computations of these systems.

Recent research has sought to assess whether large language models (LLMs) exhibit linguistic capacities comparable to those of humans. Scholars have applied methods from syntax, morphology, and semantics to analyze LLM behavior, treating these systems as novel, albeit artificial, language users. The aim is to explore whether the statistical patterns learned during training align with structured linguistic knowledge similar to that found in humans.

This approach raises concerns about anthropomorphism: attributing human-like cognitive processes to systems that operate through fundamentally different mechanisms. Scholars such as Bender et al. (2021) caution that such assumptions may bias research by framing LLMs as psycholinguistic subjects, potentially obscuring their true nature as statistical models.

Even so, the statistical generalizations formed during training may partially reflect human linguistic categories, such as syntax or morphology. While LLMs do not construct language representations in the same way humans do, their internal processes can still produce functionally similar outcomes. Thus, although traditional linguistic frameworks cannot be applied wholesale, carefully adapted methods from these disciplines may still offer valuable insights into how these systems represent and process language.

The aim of this chapter is to give an overview of the current trend of explainability, that is, the discipline that aims to uncover how language models work. To do so, the chapter begins by summarizing the most used approaches towards interpreting the LMs' inner workings (Section 2), it continues by exploring some of the key papers on syntactic (Section 3) and semantic (Section 4) explainability, and concludes by pointing out the current and future lines on language modelling to capture linguistic features (Section 5).

2. Explainability in language models

In this section of the chapter, we will analyze several research lines that attempt to discern up to what point large language models are capable of a linguistic generalization ability like that of humans. We will revise the different methodologies currently in place to discover—or dispute—the presupposed knowledge that LMs have of human language, as well as the general conclusions reached by some of these researchers. The overlook that we present in this chapter is not meant to be either exhaustive nor complete, but merely illustrative, and intended to give the reader a general feeling of the type of studies being developed in this area.

The study of linguistic knowledge falls under the supervision of a broader research area called 'interpretability,' which attempts to *interpret* the sort of general knowledge LMs possess that helps them give accurate answers (Zhao et al. 2024). This is a field that has sometimes been known as 'BERTology,' in reference to the popular language model BERT (Devlin et al., 2019), and to the name of a foundational interpretability study presented by Rogers et al. (2021). As with language, Transformer models acquire their knowledge of the world only from the texts they see during training. As such, the study of knowledge in LMs could be easily confused, and even overlap, with the study of semantics in these models; nevertheless, the former studies tend to focus more on analyzing the type and amount of encyclopedic-type knowledge of the world these models encode, as well as their reasoning of day-to-day activities (called "commonsense" knowledge). Since both general and linguistic knowledge can be studied in LMs using similar approaches—e.g., by analyzing the model's internals, it is possible that some studies that focus on world knowledge might have key ideas to be reused when analyzing linguistic knowledge—and vice versa. However, in this chapter, we will focus on those research lines that express their interest explicitly in the analysis of language comprehension only.

2.1 Prompting

One of the most straightforward ways of attempting to discern the linguistic knowledge of a language model is to ask it directly about a language phenomenon, via prompting. As suggested by its name, it makes use of prompts that are posed to a model that is to be tested. These prompts contain the explanation of a linguistically motivated task that they wish the model to solve, optionally followed by a series of examples that gear the system towards a more accurate answer, influencing their 'chain-of-thought' (Blevins et al., 2023; Di Marco, 2023). The model is then evaluated on how well it has solved that prompt. This task is mostly reserved for chatbot applications, such as ChatGPT. These tasks treat chat-based LMs as psycholinguistic subjects and expect the model to provide answers to a posed language problem with a degree of explanation similar to metalinguistic reasoning in humans. These tasks usually concern ambiguous examples (Ortega-Martin et al. 2023), or minimal pairs, such as explaining ambiguity in garden-path sentences (Amouyal et al. 2025) or choosing the correct part of speech tag for a verb or noun in a sentence (Blevins et al. 2023). There are many recent studies based on this paradigm, as it allows for an immediate and qualitative analysis of the linguistic capabilities of language models.[2] It also presents low technical overhead, as it only requires the collection of enough examples to test a specific linguistic phenomenon.

There are a few drawbacks, however, that might make this approach of study unsuitable. Many experiments, for instance, may conclude as proof of a lack of language understanding in a model that of an incorrect answer given to a linguistic question; however, failure in answering a prompt does not necessarily imply that the model has not internally encoded that linguistic construct. This is an issue addressed by Hu and Levy (2023), who perform a series of experiments on various syntactic and semantic phenomena that aim to raise some questions on the validity of these methods. On the other hand, another issue regarding this methodology is its lack of exploration of the internal model's representations. As stated above, a Transformer-based model is a statistical model, and to perform its operations, it relies on a series of transformations between vectors—representing individual tokens or sections of an input text—and a set of predefined internal matrices. The final text output of an LM—i.e., an answer to a prompt in a chatbot—is the interpretation of a series of output vector values. Commercial Transformer models, such as ChatGPT or Gemini, which are used for many of these linguistic studies, are proprietary software platforms that perform these operations behind closed servers. This means the researcher may have no access to the state of these internal computations in many current chat-based applications—although not all of them, as we will see shortly—that lead to the obtained observations. This internal data, while difficult to interpret, can prove interesting to study to understand a model's final answer. The reasons for the non-availability of this information in some widespread systems, such as ChatGPT, lie mainly in competitive reasons—companies are uneager to share their internal parameters due to risks of losing business advantage, as well as privacy concerns: some LMs are trained on potentially sensitive and private data, which have been proved to be at least partially recoverable via the analysis of these internal parameters (Melis et al., 2019; Zhu et al., 2019; Zhang et al., 2024). The closed nature of these systems also makes it impossible to discern their real capabilities, as in some cases they are released to the public with no

[2] For an in depth-analysis of prompting methods see Leidinger et al. (2023).

information on how they are internally implemented[3] or what strategies they may use to achieve better results, which may come at the expense of worse metalinguistic judgments. Opposed to these more well-known, but ultimately closed systems, there are equally performant language models that make all their internal parameters available for the end user, and which can be more beneficial for research. These include Llama (Touvron et al., 2023), released by Meta (formerly Facebook), or BLOOM (Le Scao et al., 2023).

The statements made above, however, are not to serve as a discouragement from the use of this prompting methodology—we just wish to warn the reader that a negative result presented by a study of this type is not necessarily conclusive of a model's actual capabilities.

2.2 Going beyond: Exploring a model's internal representations

Another line of study tries to venture deeper into a model's internal architecture and attempt to discern its linguistic competence from within. In general, it attempts to do so by analyzing the model's parameters—when these are available—and experimenting on how the latter influence the internal computations of the Transformer when responding to specific linguistic inputs. The analyzed components can either be the internal parameters of the models *per se* (for instance, the matrices used by the self-attention mechanism) or the intermediate representations that are obtained by these models when processing specific examples. In these methods, the model is taken as-is, without any modifications made to it, in order to analyze and gather conclusions on its internal workings. Other studies do the same but do not rely on preexisting models: they instead analyze a brand-new model that they have trained from scratch, usually on data that contains some specific language phenomena. For example, Huebner et al. (2021) examine their own custom LM, which they call BabyBERTa, trained on a dataset containing speech utterances like those of children aged one to five. Other studies also focus on interpreting a model's parameters while it is being trained, as an attempt to discover the process by which it acquires its human language skills. Examples of this approach include Chang and Bergen (2022) and Choshen et al. (2021).

---

[3] In the case of ChatGPT, it is unknown if it is a single, monolithic model being used, or a combination of models that are arranged and ranked to give the best possible answer (also called Mixture of Experts).

We will see two approaches that attempt to investigate a model's internal representations: fine-tuning and probing.

2.3 Fine-tuning

One possible method for linguistic discovery in this line is that of fine-tuning a model on linguistically oriented datasets. What does all that mean? Fine-tuning is simply a technique used to leverage an existing model and specialize it to perform a specific task (e. g. classifying texts). It consists of further training an already pretrained model with a set of examples that have been manually annotated by humans. These examples are representative of the end task the model is meant to perform. Unlike pretraining, fine-tuning requires much less data to achieve good results, and is usually able to gain excellent results thanks to the combination of this annotated data alongside the knowledge the model already possesses. The methodology for linguistic discovery, in this case, is to fine-tune a model on tasks that involve detecting specific language phenomena in these models, as a test of their linguistic abilities. In this sense, it is not too different from prompting, although the results here can be expected to be more exact due to the task directly accessing and modifying the internal parameters of a model. An example of this type of research is that of Madabushi et al. (2022). They fine-tune an English Transformer model on a task intended to discover the model's ability to choose a correct determiner article in a set of sentences.

The use of fine-tuning, however, might not be too indicative either of a model's actual linguistic performance. This is because many research lines (such as Luo et al. 2023) have demonstrated that fine-tuning, in general, might actually 'erase' or 'destroy' previously learnt knowledge—especially in the latter sections of the model (Kovaleva et al. 2019)—in an attempt to over-specialize the model in achieving a target task. Although it is likely that a well-performing linguistic task might not be able to arrive at any results at all if it were not for the original model providing a rich knowledge of grammar, it is nevertheless troublesome to assume that the discovery of this linguistic knowledge may involve a possible degradation of it and a subsequent loss of possible information. As such, this methodology, while useful, should be taken with some consideration.

In addition, both the prompting and fine-tuning paradigms might be inadequate overall for linguistic knowledge discovery because, under these methodologies, models have been reported to partially rely on *shortcuts* or shallow heuristics—such as looking in sentences at word order only—to achieve their results with less computation involved, which hurts the model's generalization abilities (Zhao et al. 2024).

2.4 Probing

There is another widespread method for linguistic discovery, on the other hand, that allows for the analysis of the model's internals without altering them. It is called the *probing* methodology. It is based on the use of probes (Conneau et al. 2018, Belinkov et al. 2019), which are classifiers trained on representations of sentences, extracted from a Transformer model, containing specific linguistic phenomena to be analyzed (e.g., correct agreement between verbs and nouns in clauses). The purpose of a probe is to attempt to learn on its own, *ex nihilo*, an underlying distribution or pattern common to a collection of these representations, one that comes to resemble the linguistic phenomenon that is wished to be studied. This pattern is not necessarily intelligible to humans, but to the model that produced it. If a probe, once trained, can consistently recognize this pattern on other sentences it is tested on, and which all have the same linguistic phenomenon, then a researcher can conclude that the model that produced these representations is encoding that linguistic information (Belinkov et al. 2019, Krasnowska-Kieraś et al. 2019). Probes are normally applied across several places in a Transformer model, to test which section of a model—called 'layers'—seems to encode that property best.[4] While probes can be tested on representations extracted from models, they can also be tested on the fixed parameters of the actual model, such as the matrices of the self-attention mechanism.[5]

The architecture of probes is not a fixed one, and it can be very varied: these modules could be any sort of system, as long as they help derive information from a model's internals. For instance, on the detection of syntax in LMs, Hewitt

[4] Though each model is likely to behave differently, there is an ongoing intuition among some researchers—not thoroughly proven, although somehow empirically reported across many papers (Zhao et al., 2024)—that lexical information seems to be (better) processed at the earlier stages of the model, while syntactic and semantic information are encoded at further steps. This fact has neither been confirmed nor dismissed at this time of writing, so it should be considered with extreme care.

[5] Rogers et al. (2021), for instance, mention a series of research studies that attempt to discover linguistic knowledge in a model's attention heads—i.e., the individual unit that makes up the self-attention mechanism, and not on the representation of a sentence as output by the model. This is done in order to see which attention head seems to encode a grammatical property best.

and Manning (2019) present a single classification layer with a simple geometric linear translation, whereas Pimentel et al. (2022) propose a more complex probe that tries to partially imitate the actual layout of the self-attention mechanism. There is a consensus in the literature, however, that probes should be as lightweight as possible (i.e., being simple classifier systems, with very few parameters). This is defended by works such as Hewitt and Liang (2019). The reason behind this is to give probes very little room to learn their task. If they were given too many parameters to learn, it might be possible that they would simply end up 'memorizing' the task they are supposed to do (Hewitt and Liang, 2019), acting as smaller versions of the original models. This would render them useless as diagnostic tools to discern a model's linguistic competence.[6]

A methodological drawback present in the probing paradigm is that it simply demonstrates the supposed existence of a linguistic phenomenon in a model's representations (Belinkov et al., 2019). That does not mean, however, that such representation might actually be in use by the model in its interpretation of language. This is an issue addressed by some researchers, such as Belinkov et al. (2019), who discuss the following dichotomy between the notions of correlation and causation:

> "From a methodological point of view, most of the relevant analysis work is concerned with correlation: how correlated are neural network components with linguistic properties? What may be lacking is a measure of causation: how does the encoding of linguistic properties affect the system output" (Belinkov et al. 2019, 52).

Some researchers acknowledge this limitation and present possible solutions that aim to proof that a probe's representations are used by a model in its understanding of language.[7] As with the general architecture in other probes,

---

[6] Hewitt and Liang (2019) tried to overcome this issue by introducing 'control probes': a type of auxiliary probes, trained alongside a set of main probes used to detect a linguistic phenomenon. These 'control' probes are purposefully misguided in their training, so that they have no clear idea of what task they are supposed to perform. If the control probe performs worse than the main probe in the same task, then the researcher can assume that no memorization of the task in the main probe has taken place.

[7] Following a proposal presented by Amini et al. (2022), we can identify these methods under a broad category called 'causal probing.' This name acknowledges these probes' attempts to provide an explanation of the *causal* relationship between the property of

the methods used to attempt to solve this task can be extremely varied. One of them, Tucker et al. (2021), working on syntax, introduces ‘counterfactual embeddings’: these consist of the sentence’s representations as produced by a model (‘embeddings’) being modified with some auxiliary data generated during the probes’ training process. Another approach, called ‘amnesic probing’ (Elazar et al., 2021), repurposes the aforementioned ‘counterfactual embeddings’ technique, in this case to remove some encoded linguistic information in a representation to see how a model acts upon that missing information. We can also list other related examples, such as information-theoretic probing (Pimentel et al. 2020) or dropout probes (Tucker et al. 2022), among others.

As a fast-developing field, the interpretability of language models is subject to emerging and brand-new techniques that are presented in the literature (e.g., in Zhao et al. 2024). In this chapter, we just mentioned a few samples of current papers, but we are looking forward to seeing how the field further develops and incorporates linguistic knowledge.

3. What do language models capture about syntax?

Probes can be used for the discovery of linguistic knowledge in many areas, such as syntax, morphology or semantics. Here, we will mention a small sample of papers investigating linguistic knowledge in LMs using probes, in the scope of syntactic knowledge, as a way of familiarizing the reader with some practical examples of this methodology. Hewitt and Manning (2019) present an experiment in which they demonstrate that a specific subspace of the vector space in the representations that are output by a model can be used to recover accurate information on several syntactic aspects between pairs of words. This includes a word’s depth in a syntactic dependency parse tree as well as distances between words in that tree. The limitations of this work lie in that it does not recover complete syntactical information on a sentence, but only gives a glimpse into a small portion of the syntactic hierarchy being encoded from raw text.

Another example of syntax is that of A. Chi et al. (2020) attempt to analyze an LM that was trained on several languages simultaneously (which is a type of model called a ‘multilingual model’). Many studies into multilingual models

---

interest and the probed model’s representations. This category stands against the other type of probes, mentioned prior, which they list under the naming of ‘correlational probes.’

attempt to see if these types of models come to encode grammatical phenomena from several different languages by using similar internal patterns. In this specific example, this is done for a series of languages including Arabic, Chinese, English and Latvian. The authors find that syntactic phenomena across multiple languages seem to be encoded in a similar manner and related locations in the model, with middle sections or 'layers' of the model performing the best. Their experiments even show the model's ability to automatically detect linguistic properties outside of the phenomena the authors explicitly set out to study, which can serve as proof of these models' generalization abilities. They warn, at the same time, however, that the model seemed to have learnt some artifacts that could be explained to it by leveraging some superficial cues from texts, such as word order.

Another multilingual example, for morphosyntax, is that of Shapiro et al. (2021), who present a probe that can automatically label words with their correct morphosyntactic features. They find that the middle sections or 'layers' of a model seem to encode this information more accurately. Other than specific layers, another research, Stańczak et al. (2023), via a custom probing solution, also discovered that there are potentially specific neurons[8] in mBERT (Devlin et al., 2019), a multilingual model that encodes shared morphosyntactic phenomena, such as number and gender, across a series of typologically different languages: Arabic, English, Finnish, Polish, Portuguese and Russian.

Beyond English and multilingual models, Guarasci et al. (2023) demonstrate, through probing as well, that monolingual BERT models for Italian are seemingly capable of encoding syntactic information well, similarly to their English counterparts, and even surpass the capabilities of dedicated Italian parsers on the null subject phenomenon and on subject-verb agreement.

Zhao and Bethard (2020) delve into a specific linguistic phenomenon, that of negation, which ranges in the areas of syntax and semantics. They do not use external probes in this case but repurpose the models' self-attention mechanism to act as a modified classifier for linguistic data. Their system presents a more complex architecture compared to other papers, despite general objections in the literature against the use of complex systems for probing. They present supporting evidence for the encoding of negation in BERT-like models. They also compare the performance of a pretrained model

---

[8] Neurons are the underlying building blocks of neural networks, akin to individual cells in living organisms.

(i.e., a base model not specialized for any specific task) against a fine-tuned model on a classification task that detects spans of words or tokens inside the scope of a negation. They do so to check whether, when fine-tuning a model, the original model's encoding of negation is either lost and replaced by an alternative mechanism or the initial representation of this phenomenon is kept by the modified model.

The papers mentioned above allude mostly to positive results, indicating good linguistic performance in these models. Some researchers, however, can also present experiments that are not so encouraging, and which can actually be proof of an LM's linguistic competence being relatively brittle. For instance, in morphosyntax and in the English language, Chaves and Richter (2021) study subject-verb agreement in sentences and find that BERT can fail to perform agreement well in some cases, such as when there are many subordinate clauses between a subject and a verb. Similarly, Wei et al. (2021) also demonstrate that subject-verb agreement rates drop when a model is exposed to less usual lexical forms. This observation is consistent with a statement made by Belinkov et al. (2019), in that "models are especially successful at capturing frequent properties, while some rare properties are more difficult to learn". In this same line, Kulnizev and Nivre (2023) also mention that "out-of-the-box models like […] Transformers dramatically fail to generalize to samples outside of their training distributions (though specialized architectures can do so trivially)".

In response to the syntactic probes of Hewitt and Manning (2019), Hall Maudslay and Cotterell (2021) also present some experiments that test probes on syntactically well-formed, yet semantically incoherent sentences, which were built by replacing some words with correctly inflected pseudo-words. Their results show performance drops, which could be an indicator of LMs not encoding syntactic phenomena separately but using an undefined combination of semantic and syntactic information for achieving their results. Similarly, Weissweiler et al. (2023), relying on a probing technique based on the Construction Grammar paradigm, show that while BERT-like models (as well as decoder-based models) in the English language are capable of detecting the comparative correlative structure in that language, they seemingly fail to grasp their meaning.

In conclusion, syntactic information seems to be partially recoverable from modern LLMs—which is a remarkable fact, considering these systems have never been exposed to any syntactic cues to interpret the texts they have seen during their training. Although this could lead the reader to assume that an LM has been able to somehow generalize syntactic knowledge—by the model only relying on the statistical cooccurrence of the individual words seen in its training texts, other lines of research have observed that this presupposed syntactic ability is sometimes prone to fail under some special conditions, such as infrequent lexical terms. This can bring up the question of whether modern LMs are truly capable of syntactic generalization akin to humans. This is an open question; however, that is left for future studies to cover.

4. Do language models have or imitate semantics?

The aim of this section is to give an overview of relevant papers that have explored different questions regarding LMs' semantic abilities. Namely, if LMs capture factual knowledge, if LMs capture semantics and concepts, if LMs do reasoning, and how they deal with figurative language.

4.1 Do language models capture factual knowledge?

One of the first works focused on elucidating whether LMs had knowledge was the one by Petroni et al. (2019) titled 'Language Models as Knowledge Bases?'. Their research question was whether they could use language models to extract factual knowledge, such as *Madrid is the Capital city of Spain.* And the claim that follows the question was: if LMs can solve this kind of question, it means they encode real-world (factual) knowledge. This would be somewhat similar to claiming that LMs can solve truth conditionals in semantics. In their work, Petroni et al. were using BERT (Devlin et al., 2019).

The model's results are promising but far from perfect. It correctly identifies the answer within the top ten options 60% of the time and within the top 100 options 80% of the time. This method allows for some flexibility as there can be multiple correct answers. Interestingly, the model performs better when it has seen the sentence's object before, but little improvement is shown when it has seen the subject in advance. Furthermore, higher prediction probabilities of the model are most correlated with the correct answer, or in other words, if the model had a very probable answer, this answer was usually the correct one, but if the model found several possible answers, its chance of failing the prediction was higher. While the model can sometimes predict real-world knowledge, it primarily relies on information it has learned from its training data. The researchers believe that future language models could replace traditional knowledge bases as a source of information. The authors of the paper claim that "this suggests that while relation extraction performance might be difficult to improve with more data, language models trained on ever-growing corpora might become a viable alternative to traditional knowledge bases extracted from text in the future." (Petroni et al. 2019, 2471).

With these results, it was hard to claim that Language Models indeed captured real-world knowledge. This would make sense as factual semantics needs real-world referents and grounding, and LMs should only be able to retrieve factual information that has already been somehow inputted to them in training in a verbalized way, and at the moment of Petroni et al.'s article, grounding was not being executed.

4.2 Do language models capture lexical semantics?

In the following years, researchers as Vúlic et al. (2020), Logan VI et al. (2022) and Pitarch et al. (2023) changed the question a bit, and proposed that although gathering real-world knowledge from LMs might still be tricky, these LMs, by being trained on such huge amounts of data, might already have enough linguistic knowledge to capture linguistic patterns and correctly make predictions on lexico-semantic tasks.

In Pitarch et al. (2023), the authors proposed to probe LMs' semantic abilities by conducting a very simple task: Lexical Relation Classification (LRC). This task consists of, given a pair of words—such as *big* and *small*—predicting the relation entailed between them, which in this case is antonymy. One can see the similarity with the work of Petroni et al., but instead of looking for factual knowledge, in this case, researchers focused only on predicting lexico-semantic relations as hyponymy, hyperonymy, antonymy, synonymy and meronymy. The claim of the paper was: if LMs encode enough linguistic knowledge, by giving it the smallest of the cues, the LM should be able to correctly predict the relation between words. As in Petroni et al., they used Pre-Trained Language Models as a base and fine-tuned them on the specific task of relation classification. The results in this case were much better, reaching above 90% of correct predictions in three out of the four tested datasets. Although promising, this is a very particular aspect of semantics, and we should not extract from it that LMs have or have not semantics, but rather that they can perform this particular task or LRC, which is in line with previous research collected for a wider range of lexico-semantic tasks in Vúlic et al. (2020).

4.3 Do language models capture concepts?

Reading Petroni et al. (2019), we could say at least there are promising pointers towards saying that LMs do encode some real-world or factual knowledge, and, in line with Vúlic et al. (2020) and Pitarch et al. (2023), we find they are quite good at capturing lexico-semantic knowledge. But can they capture concepts?[9]

In the first models that aimed at representing words as vectors (e.g., Word2Vec by Mikolov et al., 2013), a sign (word) was represented by a vector. In this type of model, polysemy is a problem because it cannot be captured. The word *book* was encoded by only a vector X, which was the same one for the word *book* in any of the following sentences: *I need to book a flight, The fantasy book is on the table,* or *Can you send me the pattern book again?* Thus, in this kind of model, it was difficult to claim that vectors were representing concepts, otherwise there should be different representations for *book* as in the sense of 'verb used to make a reservation,' 'set of written pages' or 'collection of patterns or images.'

However, at this point, it could not be a round "no, models like Word2Vec (also called static embedding models) do not capture semantics or concepts." In fact, they do capture some relevant properties and are great tools for, e.g., detecting semantic change or representing polysemy by composition (Boleda, 2020).

Then, in 2018, with BERT (Devlin et al., 2019), the question was raised again for this other family of models for representing language, called contextual embedding models. These models were able not only to assign a set of values (embedding) to represent a word, but also to update such values with respect to the context and position of the word. In such models, *book* in the sentence *I need to book a flight* is assigned a different embedding than *book* in the sentence *Can you lend me the pattern book?* Again, the interpretability of these embeddings is tricky.

Although the embedding associated with the word is now more contextually tuned, it is still not transparent. The numbers in the embedding do not represent grammatical number, concreteness, valence, or any other semantic property. They only represent which words are usually used in a given context.

In Peng et al. (2022), they ask the specific question of whether LMs encode conceptual knowledge. To answer this, they use the model to predict things like *Which is the most conceptually similar item to Inter Milan* and they provide the model with four options from which it has to decide: *Milan, Milan Fashion Week, Pohang Steelers* and *Series A* (the correct prediction would be *Pohang Steelers* since both are football clubs, no matter if the other options are lexically similar). The second test they run is testing the model to see if it detects which properties make two concepts related, given a chain of concepts such as

---

[9] Of course, if defining semantics is tricky, so is defining concepts.

animal→mammal→horse, the model needs to delimit where the attribution 'raise their young on milk' stops:

- Animals raise their young on milk → False
- Mammals raise their young on milk → True
- Horses raise their young on milk → True

Their third and last probe is to see if LMs can correctly conceptualize entities in contexts. To do so, again, they provide the model with one sentence that must be completed with one of three options (where *animal* is the correct one[10]):

Sentence: Choose the concept that best fits the context for Dolly according to the context: *Dolly is running on the grassland.* Dolly is a(n):

- animal
- mammal
- horse

After the three experiments, they conclude that LMs do encode some kind of conceptual knowledge, as results were better than random baselines, but that even after tuning the models for the particular task, results were quite low (almost for all tasks and with all the models tested below 50% of correct predictions). Thus, they point to the lack of LMs on conceptual knowledge.

4.4 Can language models do reasoning?

As studied in cognitive linguistics (Lambon Ralph et al. 2017), not only is conceptual knowledge relevant in semantics and making sense of things, but also in reasoning and being able to make inferences from previous information. Thus, it is not surprising that a wide number of papers on interpreting the semantic abilities of LMs have focused on the task of entailment. Entailment is defined as the semantic relationship between two sentences, where information from the second sentence can be inferred from the first one. Thus, when conducting this task, LMs need to reason some shared and implicit

---

[10] While in the paper the authors claim the hypernym animal is the correct answer as it can cover a broader range of animals, including mammals and non-mammals as ostriches; other authors could claim this is erroneous since this claim would also include ants or octopus which could not appear in the context 'running in the grassland,' making horses the most correct answer since they fit the context.

knowledge between these two sentences. For example, from the first sentence *My son is 13 years old* it can be inferred that *I have a son.*

Two key and recent studies on entailment are the ones by Merrill et al. (2022) and Zhang et al. (2024). Merrill et al. (2022) make a strong claim, positioning themselves towards the idea that LMs encode semantics, even if no additional grounding is given. They sustain their claim by firstly creating a controlled artificial setting that they call 'Gricean speaker'. This consists of taking the Gricean idealistic assumption that speakers try to be truthful, concise and informative (Grice, 1975). They train a toy n-gram model (Word2Vec-like but more controlled and small) with a set of possible worlds (e.g., {'John has no cats,' 'John has one cat,' 'John has two cats' and 'John has three cats'}). Then they query the model with an utterance, as *John has at least two cats,* and the model needs to predict the probability of entailment between this sentence and any of the possible worlds that were used to train the model. In their experiments, results were outstanding, and the model was able to correctly predict entailment with extremely high accuracy, especially with shorter sentences. However, one must remember this was a very controlled environment based on count-based static embeddings, and it might not be applicable to contextual (BERT-like) or generative (GPT-like) models. Also, as the authors mention—and as exhaustively revised by Robin Lakoff (1995)— Gricean semantics has its limitations and doesn't fully capture human language.

A more recent approach towards probing entailment in generative LMs (GPT-3.5, -4 and LLAMA) can be found in Zhang et al. (2024). Going back to the question of whether LMs have semantics or just imitate semantic knowledge by learning distributional patterns of words, the authors probe the model with sentences that are very similar in form but have very different semantics. Their set of example sentences is:

a) 'I was **so** certain **that** I saw you' (where there is no 'because I was certain I saw you' entailment)

b) 'I was **so** happy **that** I was freed' (where there exists entailment with 'because I was freed, I was happy')

c) 'I was **so** happy **that** I cried' (where there is also entailment of causality but in the opposite direction of the constituents 'because I was happy, I cried').

The authors create a dataset with this kind of sentence and ask the model to predict if there is entailment between pairs of them or not, and the model obtains very poor results. The authors conclude that while LMs might seem to function well in other entailment probes, they might just be relying too much on form, or in other words, they might just be learning what word patterns are usually used in entailment contexts.

4.5 How do language models deal with figurative language?

Another trending challenge in computational semantics is dealing with figurative language, especially in examples dealing with metonymy (e.g., *The government has made a decision*, where government stands for the people who make up the government) and metaphor (where one conceptual domain is used to refer to another conceptual domain, as in *Cancer is a battle*). Promising results towards the claim that LMs encode some figurative language knowledge are found in Aghazadeh et al. (2022), where they test LMs' accuracy when predicting whether a word used in a sentence is metaphoric or not. However, Pedinotti et al. (2021), when probing LMs for metaphors in a more complex and finer-graded setting, obtain poorer results. The great mismatch between the obtained results comes from the differences in framing the task. While Aghazadeh et al. use previous datasets that focus on finding whether a word is metaphoric or not in an utterance, Pedinotti et al. create a novel dataset that not only finds word-level metaphors but also genitive, attributive and complex metaphors. For each example, they also annotate whether it is a conventional metaphor, a creative metaphor, a literal expression or a nonsense expression, and so they train a model to predict the degree of conventionality of the different utterances. In a second test, for each metaphoric sentence, they provide a literal explanation and an explanation of the metaphoric sentence's reading:

a) original sentence: *The flowers nodded with the wind*

b) literal reading: 'assent head greeting'

c) metaphorical reading: 'movement wave sway'

Afterwards, they compare BERT representations for the literal and metaphorical readings of the target sentences. Their results point to LMs capturing some figurative language properties but also to the need for future work towards finding ways to improve LMs' abilities.

5. Conclusion

Overall, there is a growing body of literature interested in discovering the kinds of linguistic knowledge modern LMs might possess, ranging from syntactic to semantic phenomena, among others, through their training in large volumes of text corpora.

Most of the papers we have mentioned throughout this review rely on the probing methodology to interpret an LM's functioning. Probing consists of checking whether a model can perform a particular task, such as Lexical Relation Classification. If a measured task obtains good predictions, then the model is said to encode the grammatical property the task is evaluating. For example, in the case of Lexical Relation Classification, the models are said to have Lexico-Semantic knowledge if they are able to predict the relation entailed between pairs of words.

However, using probing to interpret LMs has its limitations, as it can only tell which tasks models can complete successfully or not, but fails at explaining why (Schwartz and Dagan 2019). Although there is a body of literature interested in demonstrating whether this information is actually in use by these models (termed as 'causal probing' by Amini et al. 2022), this is still an ongoing research area. Hence, looking at other methods to interpret the inner workings of LMs is highly advised.[11]

Further, when claiming if models have or not some properties it is also relevant to have in mind the immense relevance of the different processes involved in the pipeline of probing LMs and that can impact their output, ranging from the dataset being used and how and by whom it was annotated (Gebru 2021; Tedeschi et al. 2023), to the dataset size (Sahlgren and Lenci 2016; Zhang et al. 2021), the hyperparameters being used (Liu and Wang 2021), the metrics used to compare the results (Tedeschi et al. 2023) or the prompt used to query and train the model (Liu et al. 2023).

Another issue to highlight is that the research described in this review either examines static embeddings as Word2Vec or contextual embedding models as BERT, while very few approaches examine generative models such as the GPT family.

In terms of linguistic diversity, although the paradigm is changing, most tests and models are still conducted in English, and one should be cautious about extrapolating conclusions to other languages.

In sum, while most of the presented works show how LMs can capture some linguistically relevant features by only being trained on word forms and

---

[11] For an extensive review on different interpretability methods see Zhao et al., 2024.

distributions, they also proof their limitations and their high reliance on lexical forms rather than on deeper conceptual groundings. Thus, the debate opened by Bender and Koller (2020) remains open. LMs, on the one hand, seem to encode morphosyntax well at a first glance, but they also might be relying too much on superficial cues (e.g., word order) to accomplish this; on the other hand, we have seen that LMs do not seem to have semantics (or at least not quite). It is promising, however, to keep on exploring how researchers continue to develop LMs and add other semantic relevant features inspired by other fields as cognitive linguistics (Chersoni et al. 2021) or from multilingual perspectives (Pitarch et al. 2024).

Bibliography

Aghazadeh, Ehsan, Mohsen Fayyaz, and Yadollah Yaghoobzadeh. 2022. "Metaphors in Pre-Trained Language Models: Probing and Generalization Across Datasets and Languages." In *Proceedings of the 60th Annual Meeting of the Association for Computational Linguistics*, Volume 1: Long Papers, 2037–2050. Association for Computational Linguistics. https://doi.org/10.18653/v1/2022.acl-long.144

Amini, Afra, Tiago Pimentel, Clara Meister, and Ryan Cotterell. 2023. "Naturalistic Causal Probing for Morpho-Syntax." *Transactions of the Association for Computational Linguistics* 11: 384–403. https://doi.org/10.1162/tacl_a_00554

Amouyal, Samuel Joseph, Aya Meltzer-Asscher, and Jonathan Berant. 2025. "When the LM Misunderstood the Human Chuckled: Analyzing Garden Path Effects in Humans and Language Models." *arXiv* preprint, arXiv:2502.09307. https://arxiv.org/abs/2502.09307

Belinkov, Yonatan, and James Glass. 2019. "Analysis Methods in Neural Language Processing: A Survey." *Transactions of the Association for Computational Linguistics* 7: 49–72. https://doi.org/10.1162/tacl_a_00254

Bender, Emily, and Alexander Koller. 2020. "Climbing Towards NLU: On Meaning, Form, and Understanding in the Age of Data." In *Proceedings of the 58th Annual Meeting of the Association for Computational Linguistics*, 5185–5198. Association for Computational Linguistics. https://doi.org/10.18653/v1/2020.acl-main.463

Bender, Emily, Timnit Gebru, Angelina McMillan-Major, and Shmargaret Shmitchell. 2021. "On the Dangers of Stochastic Parrots: Can Language Models Be Too Big?" In *Proceedings of the 2021 ACM Conference on Fairness, Accountability, and Transparency (FAccT '21)*, 610–623. https://doi.org/10.1145/3442188.3445922

Blevins, Terra, Hila Gonen, and Luke Zettlemoyer. 2023. "Prompting Language Models for Linguistic Structure." In *Proceedings of the 61st Annual Meeting of the Association for Computational Linguistics*, Volume 1: Long Papers, edited by Anna Rogers, Jordan Boyd-Graber, and Naoaki Okazaki. Association for Computational Linguistics. https://doi.org/10.18653/v1/2023.acl-long.367

Boleda, Gemma. 2020. "Distributional Semantics and Linguistic Theory." *Annual Review of Linguistics* 6: 213–234. https://doi.org/10.1146/annurev-linguistics-011619-030303

Chaves, Rui P., and Stephanie N. Richter. 2021. "Look at That! BERT Can Be Easily Distracted from Paying Attention to Morphosyntax." In *Proceedings of the Society for Computation in Linguistics 2021*, edited by Allyson Ettinger, Ellie Pavlick, and Brandon Prickett. Association for Computational Linguistics. https://aclanthology.org/2021.scil-1.3

Chang, Tyler A., and Benjamin K. Bergen. 2022. "Word Acquisition in Neural Language Models." *Transactions of the Association for Computational Linguistics* 10: 1–16. https://doi.org/10.1162/tacl_a_00444

Chersoni, Emmanuele, Enrico Santus, Chu-Ren Huang, and Alessandro Lenci. 2021. "Decoding Word Embeddings with Brain-Based Semantic Features." *Computational Linguistics* 47 (3): 663–698. https://doi.org/10.1162/coli_a_00412

Chi, Ethan A., John Hewitt, and Christopher D. Manning. 2020. "Finding Universal Grammatical Relations in Multilingual BERT." In *Proceedings of the 58th Annual Meeting of the Association for Computational Linguistics*, edited by Dan Jurafsky, Joyce Chai, Natalie Schluter, and Joel Tetreault. Association for Computational Linguistics. https://doi.org/10.18653/v1/2020.acl-main.493

Choshen, Leshem, Guy Hacohen, Daphna Weinshall, and Omri Abend. 2021. "The Grammar-Learning Trajectories of Neural Language Models." In *Proceedings of the 60th Annual Meeting of the Association for Computational Linguistics*, Volume 1: Long Papers, edited by Smaranda Muresan, Preslav Nakov, and Aline Villavicencio, 8281–8297. Association for Computational Linguistics. https://doi.org/10.18653/v1/2022.acl-long.568

Conneau, Alexis, German Kruszewski, Guillaume Lample, et al. 2018. "What You Can Cram into a Single $&!#* Vector: Probing Sentence Embeddings for Linguistic Properties." In *Proceedings of the 56th Annual Meeting of the Association for Computational Linguistics*, Volume 1: Long Papers, 2126–2136. Association for Computational Linguistics. https://doi.org/10.18653/v1/P18-1198

Devlin, Jacob, Ming-Wei Chang, Kenton Lee, and Kristina Toutanova. 2019. "BERT: Pre-Training of Deep Bidirectional Transformers for Language Understanding." In *Proceedings of the 2019 Conference of the North American Chapter of the Association for Computational Linguistics: Human Language Technologies*, Volume 1 (Long and Short Papers), edited by Jill Burstein, Christy Doran, and Thamar Solorio. Association for Computational Linguistics. https://doi.org/10.18653/v1/N19-1423

Di Marco, Marion, Katharina Hämmerl, and Alexander Fraser. 2023. "Prompting Language Models for Linguistic Structure." In *Proceedings of the 2023 Conference on Empirical Methods in Natural Language Processing*, edited by Houda Bouamor, Juan Pino, and Kalika Bali. Association for

Computational Linguistics. https://doi.org/10.18653/v1/2023.emnlp-main.454

Elazar, Yanai, Shauli Ravfogel, Alon Jacovi, and Yoav Goldberg. 2021. "Amnesic Probing: Behavioral Explanation with Amnesic Counterfactuals." *Transactions of the Association for Computational Linguistics* 9: 160–175. https://doi.org/10.1162/tacl_a_00359.

Gebru, Timnit, Jamie Morgenstern, Briana Vecchione, et al. 2021. "Datasheets for Datasets." *Communications of the ACM* 64 (12): 86–92. https://doi.org/10.1145/3458723

Grice, H. Paul. 1975. "Logic and Conversation." In *Syntax and Semantics, Volume 3: Speech Acts*, edited by Peter Cole and Jerry L. Morgan. New York: Academic Press.

Guarasci, Raffaele, Stefano Silvestri, Giuseppe De Pietro, Hamido Fujita, and Massimo Esposito. 2023. "Assessing BERT's Ability to Learn Italian Syntax: A Study on Null-Subject and Agreement Phenomena." *Journal of Ambient Intelligence and Humanized Computing* 14: 289–303. https://doi.org/10.1007/s12652-021-03297-4

Hall Maudslay, Rowan, and Ryan Cotterell. 2021. "Do Syntactic Probes Probe Syntax? Experiments with Jabberwocky Probing." In *Proceedings of the 2021 Conference of the North American Chapter of the Association for Computational Linguistics: Human Language Technologies*, edited by Kristina Toutanova et al. Association for Computational Linguistics. https://doi.org/10.18653/v1/2021.naacl-main.11

Hewitt, John, and Percy Liang. 2019. "Designing and Interpreting Probes with Control Tasks." In *Proceedings of the 2019 Conference on Empirical Methods in Natural Language Processing and the 9th International Joint Conference on Natural Language Processing (EMNLP-IJCNLP)*, edited by Kentaro Inui, Jing Jiang, Vincent Ng, and Xiaojun Wan. Association for Computational Linguistics. https://doi.org/10.18653/v1/D19-1275

Hewitt, John, and Christopher D. Manning. 2019. "A Structural Probe for Finding Syntax in Word Representations." In *Proceedings of the 2019 Conference of the North American Chapter of the Association for Computational Linguistics: Human Language Technologies*, Volume 1 (Long and Short Papers), edited by Jill Burstein, Christy Doran, and Thamar Solorio. Association for Computational Linguistics. https://doi.org/10.18653/v1/N19-1419

Hu, Jennifer, and Roger Levy. 2023. "Prompting Is Not a Substitute for Probability Measurements in Large Language Models." In *Proceedings of the 2023 Conference on Empirical Methods in Natural Language Processing*, edited by Houda Bouamor, Juan Pino, and Kalika Bali. Association for Computational Linguistics. https://doi.org/10.18653/v1/2023.emnlp-main.306

Huebner, Philip A., Elior Sulem, Cynthia Fisher, and Dan Roth. 2021. "BabyBERTa: Learning More Grammar with Small-Scale Child-Directed Language." In *Proceedings of the 25th Conference on Computational Natural Language Learning*, edited by Arianna Bisazza and Omri Abend. Association for Computational Linguistics. https://doi.org/10.18653/v1/2021.conll-1.49

Kovaleva, Olga, Alexey Romanov, Anna Rogers, and Anna Rumshisky. 2019. "Revealing the Dark Secrets of BERT." In *Proceedings of the 2019 Conference on Empirical Methods in Natural Language Processing and the 9th International Joint Conference on Natural Language Processing (EMNLP-IJCNLP)*, edited by Kentaro Inui, Jing Jiang, Vincent Ng, and Xiaojun Wan. Association for Computational Linguistics. https://doi.org/10.18653/v1/D19-1445

Krasnowska-Kieraś, Katarzyna, and Alina Wróblewska. 2019. "Empirical Linguistic Study of Sentence Embeddings." In *Proceedings of the 57th Annual Meeting of the Association for Computational Linguistics*, edited by Anna Korhonen, David Traum, and Lluís Màrquez. Association for Computational Linguistics. https://doi.org/10.18653/v1/P19-1573

Kulmizev, Artur, and Joakim Nivre. 2022. "Schrödinger's Tree—On Syntax and Neural Language Models." *Frontiers in Artificial Intelligence* 5: 796788. https://doi.org/10.3389/frai.2022.796788

Lakoff, Robin T. 1995. "Conversational Logic." In *Handbook of Pragmatics*, edited by Jef Verschueren, Jan-Ola Östman, and Jan Blommaert. https://doi.org/10.1075/hop.m.con5

Le Scao, Teven, Angela Fan, Christopher Akiki, et al. 2023. "BLOOM: A 176B-Parameter Open-Access Multilingual Language Model." *arXiv* preprint, arXiv:2211.05100. https://arxiv.org/abs/2211.05100

Leidinger, Alina, Robert van Rooij, and Ekaterina Shutova. 2023. "The Language of Prompting: What Linguistic Properties Make a Prompt Successful?" In *Findings of the Association for Computational Linguistics: EMNLP 2023*, 9210–9232. https://doi.org/10.18653/v1/2023.findings-emnlp.618

Liu, Pengfei, Weizhe Yuan, Jinlan Fu, Zhengbao Jiang, Hiroaki Hayashi, and Graham Neubig. 2023. "Pre-Train, Prompt, and Predict: A Systematic Survey of Prompting Methods in Natural Language Processing." *ACM Computing Surveys* 55 (9): 1–35. https://doi.org/10.1145/3560815

Liu, Xueqing, and Chi Wang. 2021. "An Empirical Study on Hyperparameter Optimization for Fine-Tuning Pre-Trained Language Models." In *Proceedings of the 59th Annual Meeting of the Association for Computational Linguistics and the 11th International Joint Conference on Natural Language Processing*, Volume 1: Long Papers, 2286–2300. https://doi.org/10.18653/v1/2021.acl-long.178

Logan IV, Robert, Ivana Balažević, Eric Wallace, Fabio Petroni, Sameer Singh, and Sebastian Riedel. 2022. "Cutting Down on Prompts and Parameters: Simple Few-Shot Learning with Language Models." In *Findings of the Association for Computational Linguistics: ACL 2022*, 2824–2835. https://doi.org/10.18653/v1/2022.findings-acl.222

Luo, Yun, Zhen Yang, Fandong Meng, Yafu Li, Jie Zhou, and Yue Zhang. 2023. "An Empirical Study of Catastrophic Forgetting in Large Language Models during Continual Fine-Tuning." *arXiv* preprint, arXiv:2308.08747. https://arxiv.org/abs/2308.08747

Madabushi, Harish Tayyar, Dagmar Divjak, and Petar Milin. 2022. "Abstraction Not Memory: BERT and the English Article System." In *Proceedings of the 2022*

Conference of the North American Chapter of the Association for Computational Linguistics: Human Language Technologies, edited by Marine Carpuat, Marie-Catherine de Marneffe, and Ivan Vladimir Meza Ruiz. Association for Computational Linguistics. https://doi.org/10.18653/v1/2022.naacl-main.67

Melis, Luca, Congzheng Song, Emiliano De Cristofaro, and Vitaly Shmatikov. 2019. "Exploiting Unintended Feature Leakage in Collaborative Learning." In *2019 IEEE Symposium on Security and Privacy (SP)*. IEEE. https://doi.org/10.1109/SP.2019.00029

Merrill, William, Alex Warstadt, and Tal Linzen. 2022. "Entailment Semantics Can Be Extracted from an Ideal Language Model." In *Proceedings of the 26th Conference on Computational Natural Language Learning (CoNLL)*, 176–193. https://doi.org/10.18653/v1/2022.conll-1.13

Mikolov, Tomas, Ilya Sutskever, Kai Chen, Greg Corrado, and Jeffrey Dean. 2013. "Distributed Representations of Words and Phrases and Their Compositionality." *Advances in Neural Information Processing Systems* 26: 3111–3119.

Ortega-Martin, Miguel, Óscar García-Sierra, Alfonso Ardoiz, et al. 2023. "Linguistic Ambiguity Analysis in ChatGPT." *arXiv* preprint, arXiv:2302.06426 . https://arxiv.org/abs/2302.06426

Pedinotti, Paolo, Eliana Di Palma, Ludovica Cerini, and Alessandro Lenci. 2021. "A Howling Success or a Working Sea? Testing What BERT Knows about Metaphors." In *Proceedings of the Fourth BlackboxNLP Workshop on Analyzing and Interpreting Neural Networks for NLP*, 192–204. https://doi.org/10.18653/v1/2021.blackboxnlp-1.13

Peng, Hao, Xiaozhi Wang, Shending Hu, et al. 2022. "COPEN: Probing Conceptual Knowledge in Pre-Trained Language Models." In *Proceedings of the 2022 Conference on Empirical Methods in Natural Language Processing*, 5015–5035. https://doi.org/10.18653/v1/2022.emnlp-main.335

Pitarch, Lucía, Jorge Bernad, Lacramioara Dranca, Carlos Bobed Lisbona, and Jorge Gracia. 2023. "No Clues Good Clues: Out of Context Lexical Relation Classification." In *Proceedings of the 61st Annual Meeting of the Association for Computational Linguistics*, Volume 1: Long Papers, 5607–5625. https://doi.org/10.18653/v1/2023.acl-long.308

Petroni, Fabio, Tim Rocktäschel, Sebastian Riedel, et al. 2019. "Language Models as Knowledge Bases?" In *Proceedings of the 2019 Conference on Empirical Methods in Natural Language Processing and the 9th International Joint Conference on Natural Language Processing*, 2463–2473. https://doi.org/10.18653/v1/D19-1250

Pimentel, Tiago, Josef Valvoda, Rowan Hall Maudslay, Ran Zmigrod, Adina Williams, and Ryan Cotterell. 2020. "Information-Theoretic Probing for Linguistic Structure." In *Proceedings of the 58th Annual Meeting of the Association for Computational Linguistics*, edited by Dan Jurafsky, Joyce Chai, Natalie Schluter, and Joel Tetreault. Association for Computational Linguistics. https://doi.org/10.18653/v1/2020.acl-main.420

Pimentel, Tiago, Josef Valvoda, Niklas Stoehr, and Ryan Cotterell. 2022. "Attentional Probe: Estimating a Module's Functional Potential." In *Proceedings of the 2022 Conference on Empirical Methods in Natural Language Processing*, edited by Yoav Goldberg, Zornitsa Kozareva, and Yue Zhang. Association for Computational Linguistics. https://doi.org/10.18653/v1/2022.emnlp-main.788

Pitarch, Lucía, Carlos Bobed Lisbona, David Abián, Jorge Gracia, and Jorge Bernad. 2024. "Building MUSCLE: A Dataset for Multilingual Semantic Classification of Links Between Entities." In *Proceedings of the 2024 Joint International Conference on Computational Linguistics, Language Resources and Evaluation (LREC-COLING 2024)*, 2580–2594

Ralph, Mathew, Elizabeth Jefferies, Karalyn Patterson, and Timothy Rogers. 2017. "The Neural and Computational Bases of Semantic Cognition." *Nature Reviews Neuroscience* 18 (1): 42–55. https://doi.org/10.1038/nrn.2016.150

Rogers, Anna, Olga Kovaleva, and Anna Rumshisky. 2021. "A Primer in BERTology: What We Know about How BERT Works." *Transactions of the Association for Computational Linguistics* 8: 842–866. https://doi.org/10.1162/tacl_a_00349

Sahlgren, Magnus, and Alessandro Lenci. 2016. "The Effects of Data Size and Frequency Range on Distributional Semantic Models." In *Proceedings of the 2016 Conference on Empirical Methods in Natural Language Processing*, 975–980. https://doi.org/10.18653/v1/D16-1099

Shapiro, Naomi, Amandalynne Paullada, and Shane Steinert-Threlkeld. 2021. "A Multilabel Approach to Morphosyntactic Probing." In *Findings of the Association for Computational Linguistics: EMNLP 2021*, edited by Marie-Francine Moens, Xuanjing Huang, Lucia Specia, and Scott Wen-tau Yih. Association for Computational Linguistics. https://doi.org/10.18653/v1/2021.findings-emnlp.382

Shwartz, Vered, and Ido Dagan. 2019. "Still a Pain in the Neck: Evaluating Text Representations on Lexical Composition." *Transactions of the Association for Computational Linguistics* 7: 403–419. https://doi.org/10.1162/tacl_a_00277

Stańczak, Karolina, Lucas Torroba Hennigen, Adina Williams, Ryan Cotterell, and Isabelle Augenstein. 2023. "A Latent-Variable Model for Intrinsic Probing." In *Proceedings of the Thirty-Seventh AAAI Conference on Artificial Intelligence and Thirty-Fifth Conference on Innovative Applications of Artificial Intelligence and Thirteenth Symposium on Educational Advances in Artificial Intelligence (AAAI'23/IAAI'23/EAAI'23)*, 13591–13599. https://doi.org/10.1609/aaai.v37i11.26593

Tedeschi, Simone, Johan Bos, Thierry Declerck, et al. 2023. "What's the Meaning of Superhuman Performance in Today's NLU?" In *Proceedings of the 61st Annual Meeting of the Association for Computational Linguistics*, Volume 1: Long Papers, 12471–12491. https://doi.org/10.18653/v1/2023.acl-long.697

Touvron, Hugo, Thibaut Lavril, Gautier Izacard, et al. 2023. "LLaMA: Open and Efficient Foundation Language Models." *arXiv* preprint, arXiv:2302.13971. https://arxiv.org/abs/2302.13971

Tucker, Mycal, Peng Qian, and Roger Levy. 2021. "What If This Modified That? Syntactic Interventions via Counterfactual Embeddings." In *Findings of the Association for Computational Linguistics: ACL-IJCNLP 2021*, edited by

Chengqing Zong, Fei Xia, Wenjie Li, and Roberto Navigli. Association for Computational Linguistics. https://doi.org/10.18653/v1/2021.findings-acl.76

Tucker, Mycal, Tiwalayo Eisape, Peng Qian, Roger Levy, and Julie Shah. 2022. "When Does Syntax Mediate Neural Language Model Performance? Evidence from Dropout Probes." In *Proceedings of the 2022 Conference of the North American Chapter of the Association for Computational Linguistics: Human Language Technologies*, edited by Marine Carpuat, Marie-Catherine de Marneffe, and Ivan Vladimir Meza Ruiz. Association for Computational Linguistics. https://doi.org/10.18653/v1/2022.naacl-main.394

Vaswani, Ashish, Noam Shazeer, Niki Parmar, et al. 2017. "Attention Is All You Need." In *Proceedings of the 31st International Conference on Neural Information Processing Systems (NeurIPS '17)*, edited by Ulrike von Luxburg, Isabelle Guyon, Samy Bengio, Hanna Wallach, and Rob Fergus. Curran Associates Inc.

Vulić, Ivan, Edoardo Maria Ponti, Robert Litschko, Goran Glavaš, and Anna Korhonen. 2020. "Probing Pretrained Language Models for Lexical Semantics." In *Proceedings of the 2020 Conference on Empirical Methods in Natural Language Processing (EMNLP)*, 7222–7240. https://doi.org/10.18653/v1/2020.emnlp-main.586

Wei, Jason, Dan Garrette, Tal Linzen, and Ellie Pavlick. 2021. "Frequency Effects on Syntactic Rule Learning in Transformers." In *Proceedings of the 2021 Conference on Empirical Methods in Natural Language Processing*, edited by Marie-Francine Moens, Xuanjing Huang, Lucia Specia, and Scott Wen-tau Yih. Association for Computational Linguistics. https://doi.org/10.18653/v1/2021.emnlp-main.72

Weissweiler, Leonie, Valentin Hofmann, Abdullatif Köksal, and Hinrich Schütze. 2022. "The Better Your Syntax, the Better Your Semantics? Probing Pretrained Language Models for the English Comparative Correlative." In *Proceedings of the 2022 Conference on Empirical Methods in Natural Language Processing*, edited by Yoav Goldberg, Zornitsa Kozareva, and Yue Zhang. Association for Computational Linguistics. https://doi.org/10.18653/v1/2022.emnlp-main.746

Zhang, Yian, Alex Warstadt, Xiaocheng Li, and Samuel R. Bowman. 2021. "When Do You Need Billions of Words of Pretraining Data?" In *Proceedings of the 59th Annual Meeting of the Association for Computational Linguistics and the 11th International Joint Conference on Natural Language Processing*, Volume 1: Long Papers, 1112–1125. https://doi.org/10.18653/v1/2021.acl-long.90

Zhang, Zhuo, Jintao Huang, Xiangjing Hu, et al. 2024. "Revisiting Data Reconstruction Attacks on Real-World Dataset for Federated Natural Language Understanding." In *Proceedings of the 2024 Joint International Conference on Computational Linguistics, Language Resources and Evaluation (LREC-COLING 2024)*, edited by Nicoletta Calzolari, Min-Yen Kan, Veronique Hoste, Alessandro Lenci, Sakriani Sakti, and Nianwen Xue. ELRA and ICCL. https://aclanthology.org/2024.lrec-main.1227

Chapter 6

ChatGPT and linguistic theory, with a focus on morphology

Stela Manova

Manova AI

Abstract: This chapter proposes to move beyond the abstract level at which the discussion of the extent to which Large Language Models (LLMs) can serve as theories of language has been conducted. Thus, the author looks at how LLMs work technically and what their internal organization means for linguistic theory. Since algorithms in different LLMs may differ, the author focuses on ChatGPT, paying particular attention to the so-called Byte Pair Encoding (BPE), a form-based subword tokenization algorithm that is at the core of ChatGPT. The author illustrates how this algorithm treats language and compares it to current research in theoretical morphology. Following the logic of the Big O notation in computer science, she argues that a form-based analysis is less complex than a semantics-based one, suggesting that ChatGPT could be a suitable model of human knowledge of morphology.

Keywords: Large Language Models, Artificial Intelligence, Morphology, Natural Language Processing, Byte Pair Encoding, Tokenization

***

1. Introduction

With the rise of Large Language Models (LLMs), in Linguistic Theory (LT), much attention has been paid to whether LLMs can serve as theories of language (Piantadosi 2023 and replies to other scholars in it). However, the discussion has been one-sided: only linguists have participated in it (Chomsky et al. 2023; Haider 2023; Katzir 2023; Moro et al. 2023; Rawski and Baumont 2023; Sauerland 2023), and consequently the approach adopted has been an abstract one, with an emphasis on semantics and syntax, though

LLMs understand and generate language based exclusively on form. The technical organization of LLMs and the question of what this organization means for LT have remained unaddressed. The goal of my contribution is to provide the missing pieces. Since algorithms in different LLMs may differ, I tackle ChatGPT.

A GPT (an abbreviation for Generative Pre-trained Transformer) is a type of LLM based on an artificial neural network (transformer architecture) and pre-trained on large data sets of unlabeled text, i.e., a GPT does not use the grammar of the type known from LT.

ChatGPT was launched by OpenAI on November 30, 2022, and, as its name indicates, it is an LLM chatbot. It has a user-friendly interface and has been additionally trained for dialogue with humans. The most surprising feature of ChatGPT from a linguistic point of view (because ChatGPT can accomplish non-linguistic tasks as well) is its ability to quickly understand prompts and generate human-like texts in a real-time chat. The chatbot is fluent in a number of languages, and it does not operate with words but with subword units (tokens) that are not associated with semantics (on meaning-form mapping in theoretical and computational approaches in morphology, see Manova et al. 2020). Roughly, in ChatGPT, semantics emerges effortlessly as an effect of the use of long sequences of tokens. The NVIDIA website provides maybe the best explanation of how form and meaning relate in an LLM:

> "A transformer model is a neural network that learns context and thus meaning by tracking relationships in sequential data like the words in this sentence".[1]

The fact that ChatGPT understands and generates language based on subword units is the explanation for why this chapter has an emphasis on morphology.

ChatGPT can accomplish complex semantic and syntactic tasks (Haider 2023), but since it is not a word-based model, it faces significant difficulties with the solution to a very basic linguistic task, such as word counting, which makes the chatbot's quickness and fluency during a real-time chat a mystery from a linguistic point of view. Another perplexing fact concerns ChatGPT's vocabulary, cl100k_base. Although ChatGPT has been trained on huge amounts of text, its vocabulary is fixed-size and contains only 100k tokens

---

[1] https://blogs.nvidia.com/blog/what-is-a-transformer-model/

(subword units).[2] The list of all ChatGPT tokens (from now on, I will refer to it as 'the List') is available from a GitHub page.[3] As I will demonstrate later, the List can be verified with the ChatGPT interactive tokenizer (from now on: 'the Tokenizer').[4]

I return to all the above-mentioned issues below. In this chapter, I demonstrate that ChatGPT's internal organization cannot be explained with a single linguistic theory but with a combination of insights from different theories. Since the focus of the text is on morphology, I pay special attention to Distributed Morphology (Halle and Marantz 1993; Harley and Noyer 1999; Embick and Noyer 2007; Bobaljik 2017) and Paradigm Function Morphology (Stump 2001, 2016; Stump and Finkel 2013; Bonami and Stump 2017). I claim that ChatGPT is compatible with LT and can serve as a theory of language.

The chapter has the following structure. In section 2, I explain and illustrate tokenization, as well as how ChatGPT works with tokens, namely that when accomplishing a task, the model processes language not orthographically but in terms of token IDs. Section 3 is devoted to the notion of complexity: I explain why form-based language processing (the case of ChatGPT) is less complex than semantics-based language processing (the case of LT). In LT of the two morphological components, word-formation and inflection, the former has always been seen as semantically-driven; therefore, in section 4, I demonstrate a form-based analysis of word-formation and provide psycholinguistic evidence that this is how native speakers process morphology. Section 5 accommodates the discussion, i.e., in this section, I explain, among other things, why ChatGPT can serve as a theory of language. In the last section, conclusions are drawn.

2. Words versus tokens

Words are the central units of analysis in LT, yet ChatGPT does not work with words but with tokens (subword units). A token can be represented as a text (letters), see the red quadrangle at the bottom of Figure 6.1, or as a token ID, which is a number, see the red quadrangle at the bottom of Figure 6.2. Thus, in

---

[2] When the paper was written only the cl100k_base vocabulary list was publicly available; currently, ChatGPT uses the o200k_base vocabulary, with 200k tokens, https://github.com/kaisugi/gpt4_vocab_list/tree/main.

[3] https://github.com/kaisugi/gpt4_vocab_list/tree/main (last accessed July 28, 2025)

[4] The interactive Tokenizer has been prepared and maintained by OpenAI and can be found on their webpage: https://platform.openai.com/tokenizer

the Tokenizer *-er*/'er' has the token ID [261], Figure 6.2. From a technical point of view, a computer works with numbers faster than with letters; therefore, in an LLM, a prompt is first tokenized (at this stage, tokens are text), then encoded into token IDs, which are numbers, and only afterwards processed. When the task stated in the prompt is solved, ChatGPT's answer is a sequence of IDs, and it is then decoded into text. ChatGPT generates sequences of forms token-by-token and predicts the next token based on the statistics of co-occurrence of tokens in training data. Unlike in LT, ChatGPT composes and parses in two different ways, e.g., *metalanguage* is generated as *meta+language* (IDs [8999, 4221]), but parsed as *metal-anguage* (IDs [9501, 2681]); see also the parsing of the token *linguistic* [65767] in *crooslinguistic*, Figures 6.10 through 6.13.

Figure 6.1: -er (token as Text)

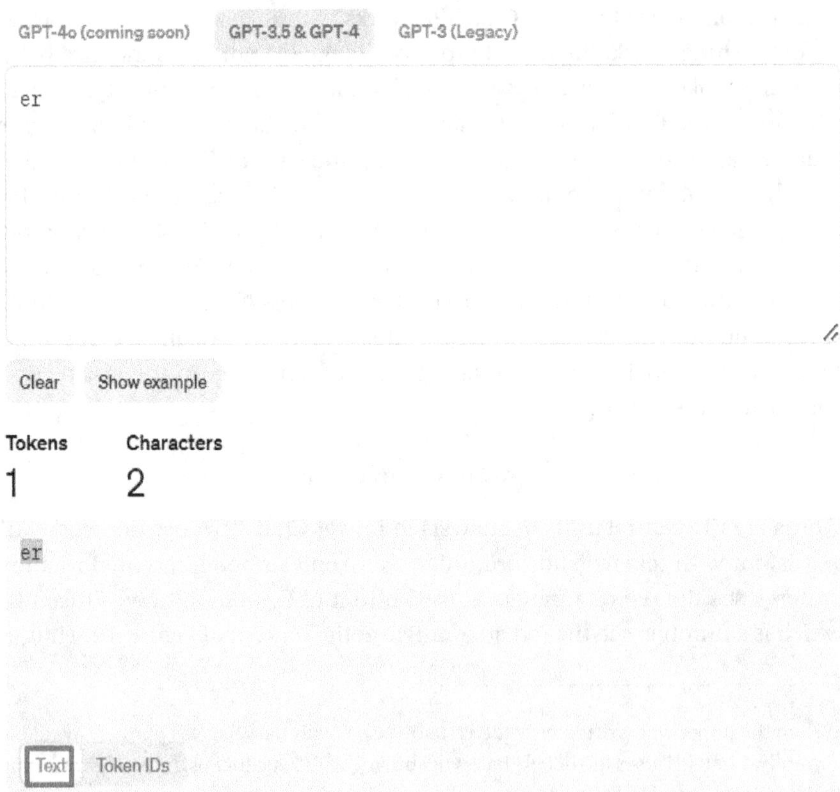

Source: https://platform.openai.com/tokenizer

Figure 6.2: -er (token ID)

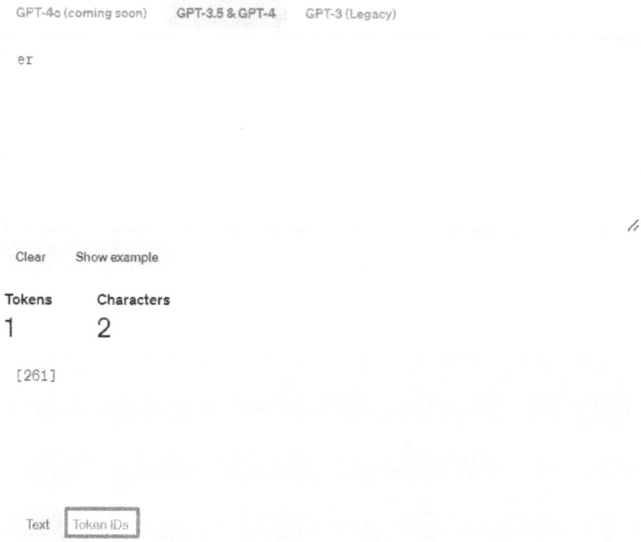

Many tokens coincide with linguistic units, as is the case with 'er' / *-er* (Figure 6.1 and Figure 6.2), which coincides with the English suffix *-er*, as in *read-er*, *open-er*, etc. As can be seen from the List, some of the tokens are letters, others are morphemes, still others are words, and there are also tokens that, according to LT, are not linguistic units, see, e.g., #257, #258, #261, and #263 in Figure 6.3 which is a screenshot of a segment extracted from the List. These four tokens are simply free spaces of different lengths.

Figure 6.3: Tokens that are not linguistic units

257	' '
258	' '
259	'in'
260	' t'
261	' '
262	'er'
263	' '

Source: https://github.com/kaisugi/gpt4_vocab_list/blob/main/cl100k_base_vocab_list.txt

For ChatGPT, a free space is a meaningful character. Therefore, 'er' / *-er* (Figures 6.1, 6.2) differs from 'er' / *er-* (Figures 6.4, 6.5), and consequently these two tokens have different IDs, [261] (Figure 6.2) and [2781] (Figure 6.5), respectively. The differentiation of the two tokens is of particular importance because [2781], with the free space before it, indicates a word beginning. This is the explanation why 'er' [261] in Figures 6.1, 6.2 consists of two characters, while 'er' [2781] in Figures 6.4, 6.5 has three characters (i.e., the free space counts as a character). A smaller ID means a token with a higher frequency, i.e., 'er' / *-er* [261] is more frequent than 'er' / *er-* [2781].

Figure 6.4: *er-* (token as text)

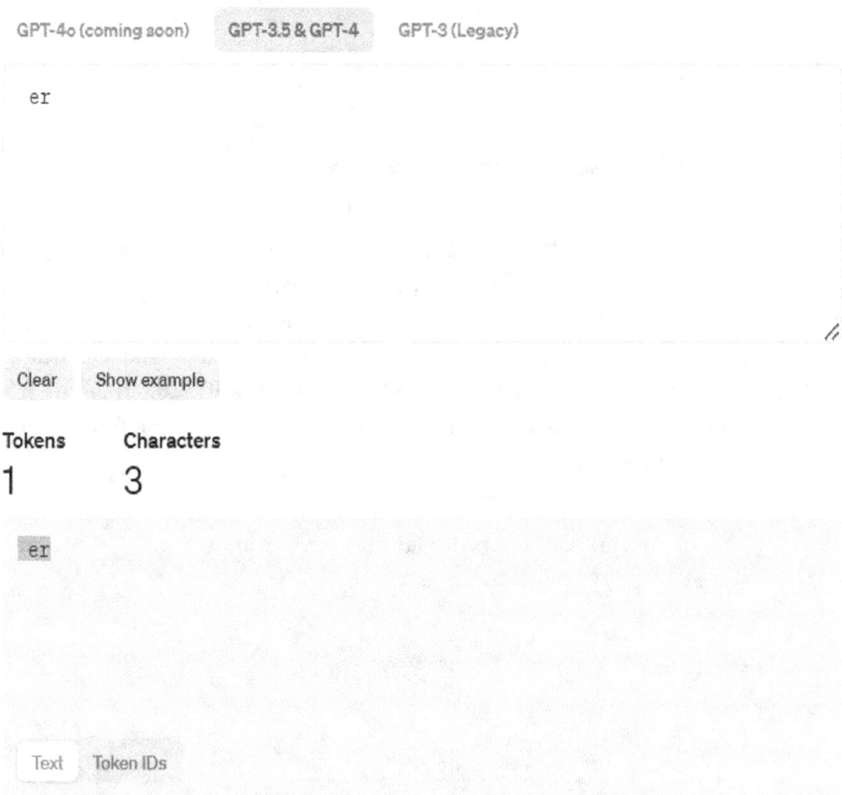

Source: https://platform.openai.com/tokenizer

Figure 6.5: *er-* (token ID)

Tokenization makes possible the representation of a large amount of text with a small set of tokens, the latter are comparable with letters, morphemes and highly frequent words in LT. Now, if tokenization is so important, why is *er-* in the List #262, but in the Tokenizer its ID is [261]. This difference results from the fact that computers, unlike humans, start counting from [0], i.e., the Tokenizer starts from zero, while the List follows the human way of counting and starts from one. Therefore, #1 in the List, '!', has the token ID [0] in the Tokenizer, Figures 6.6 and 6.7, respectively.

Figure 6.6: Token #1 in the List

1	'!'
2	'"'
3	'#'
4	'$'
5	'%'

Figure 6.7: Token ID [0] in the Tokenizer

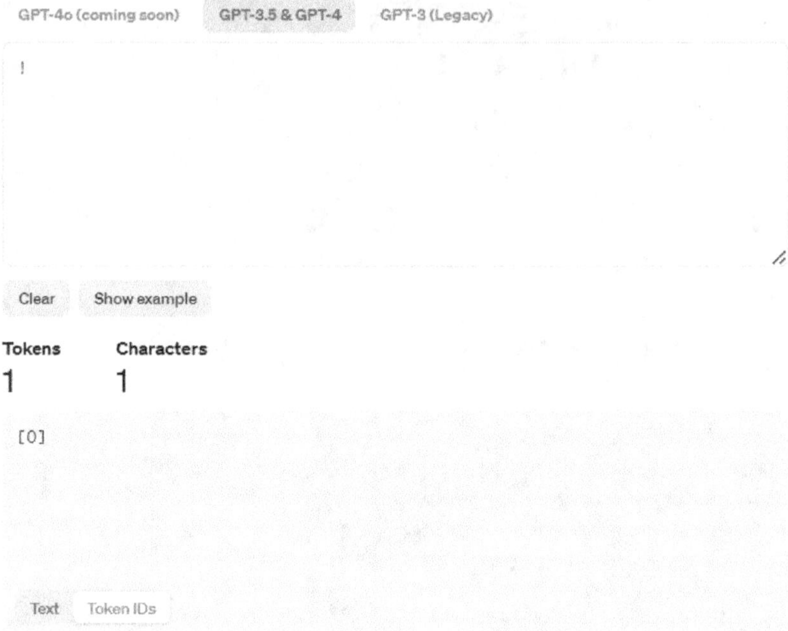

Regarding vocabulary length, ChatGPT's vocabulary is declared to be of 100k tokens, yet the List contains 100256 entries. This is because we need items for vocabulary building, i.e., the first 256 items serve for initialization (building of the rest) of the vocabulary.

For tokenization, ChatGPT uses tiktoken, a fast Byte Pair Encoding (BPE) algorithm (Sennrich et al. 2016).[5] The most important steps of the algorithm, extracted from the explanation at this link, are listed in (1). For the sake of clarity, I expanded the descriptions of some of the algorithm's steps with additional information.

Now let me illustrate tokenization in practice: Imagine that we have a text corpus that should be tokenized and that this corpus consists of four words, as shown in (1):

(1) Byte Pair Encoding (BPE)

---

[5] https://github.com/openai/tiktoken. An easy-to-understand explanation of the BPE logic can be found at: https://www.geeksforgeeks.org/byte-pair-encoding-bpe-in-nlp/

Text corpus: "**ab**", "**bc**", "**bcd**", "**cde**" (i.e., the corpus consists of four words)

a) Step 1: Initialize the vocabulary (i.e., establish all characters (bytes) used in the corpus)

   Vocabulary = {"a", "b", "c", "d", "e"}

b) Step 2: Calculate the frequency of each character

   Frequency = {"a": 1, "b": 2, "c": 3, "d": 2, "e": 1}

c) Step 3a: Find the most frequent pair of two characters (in the same position)

   The most frequent pair in the initial position is "bc", with a frequency of 2. ("cd" does not count because in "bcd" it is in a final position, while in "cde", it is in an initial position)

d) Step 3b: Merge the pair

   Merge "b" and "c" to create a new subword unit "bc".

e) Step 3c: Update frequency

   Update the frequency count of "bc":

   Frequency = {"a": 1, "b": 2, "c": 3, "d": 2, "e": 1, "bc": 2}

f) Step 3d: Add the new subword unit to the vocabulary

   Add "bc" to the vocabulary:

   Vocabulary = {"a", "b", "c", "d", "e", "bc"}

   Repeat steps 3a-3d until the desired vocabulary size is reached.[6]

As already mentioned, the ChatGPT vocabulary is limited to 100k. In other words, the frequency of the 100k-th token serves as a threshold for vocabulary inclusion. All items that are part of the vocabulary count as single tokens and are therefore considered non-derived, which makes 'linguistic', as in *linguistic research*, a single token, i.e., a non-derived item (*research* is also a single token), Figures 6.8, 6.9; but 'linguistic' in *crosslinguistic* is a derived item, Figures 6.10,

[6] A Python implementation of the BPE can be found at the bottom of: https://www.geeksforgeeks.org/byte-pair-encoding-bpe-in-nlp/. Further useful information on tokenization at: https://cookbook.openai.com/examples/how_to_count_tokens_with_tiktoken. An alternative Tokenizer is available as a web app at: https://tiktokenizer.vercel.app/. With this app, the reader can experiment with the tokenization strategies employed by different models.

6.11. For the sake of completeness, Figures 6.12, 6.13 demonstrate that spelling (hyphenated versus together) does not influence the parsing of 'linguistic.'

Figure 6.8: 'linguistic research' (Text)

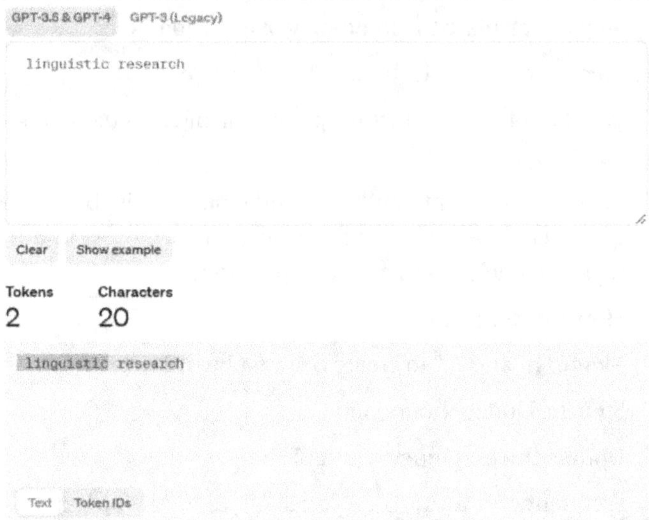

Figure 6.9: 'linguistic research' (IDs)

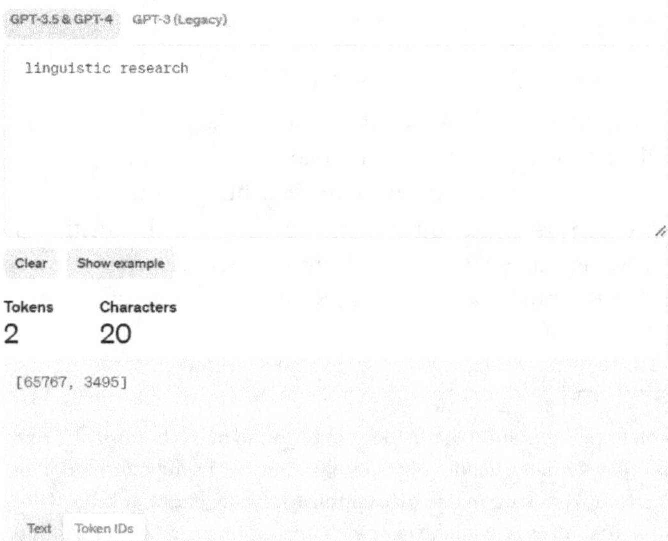

Figure 6.10: 'crosslinguistic' (Text)

Figure 6.11: 'crosslinguistic' (IDs)

Figure 6.12: 'cross-linguistic" (Text)

Figure 6.13: 'cross-linguistic" (IDs)

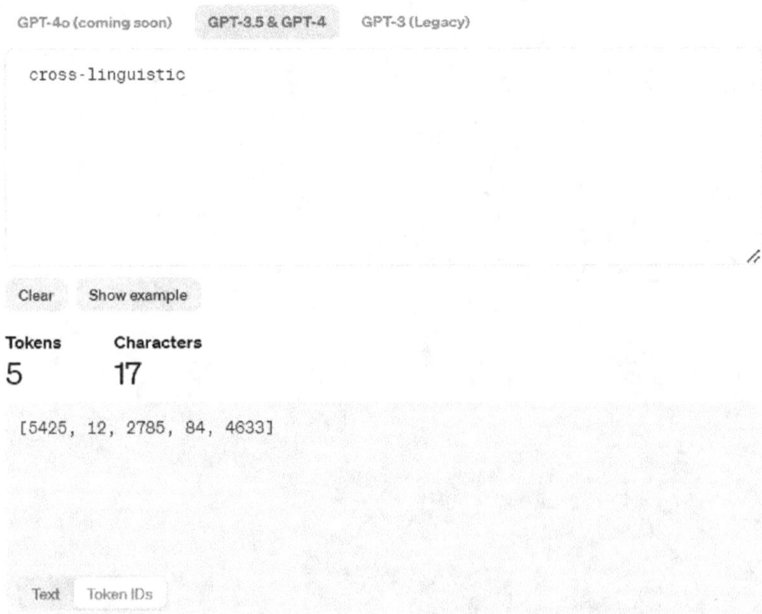

Because of the hyphen, see [12] in Figure 6.13, *crosslinguistic* and *cross-linguistic* differ in a character, 16 and 17 characters, respectively, Figures 6.10, 6.11 and Figures 6.12, 6.13.

In sum, ChatGPT works with tokens (subword units), the latter may coincide in form with letters, morphemes or highly frequent words; there are also tokens that do not coincide with linguistic units. Tokens are not associated with semantics. Language is a long, uninterrupted sequence of tokens in which spaces indicate word beginnings. Depending on position and frequency of use in this position, the same sequence of letters may have different IDs or may be treated as both non-derived and derived. A smaller ID indicates a higher frequency. ChatGPT does not process text but IDs.

I am aware that for a linguist, it is somehow hard to believe that ChatGPT does not work with words but with tokens. So, as an illustration of the issue, I gave the task in (2) to ChatGPT. I asked the chatbot to count repeated words in a text (pay attention to the length of the prompt, namely that it is not a single word but a sentence). The text used for word counting is the abstract of one of the versions of Piantadosi (2023).

(2) Counting repeated words with ChatGPT:

> **Prompt:** List and count words used more than once in the following text:
>
> "The rise and success of large language models undermines virtually every strong claim for the innateness of language that has been proposed by generative linguistics. Modern machine learning has subverted and bypassed the entire theoretical framework of Chomsky's approach, including its core claims to particular insights, principles, structures, and processes. I describe the sense in which modern language models implement genuine theories of language, including representations of syntactic and semantic structure. I highlight the relationship between contemporary models and prior approaches in linguistics, namely those based on gradient computations and memorized constructions. I also respond to several critiques of large language models, including claims that they can't answer 'why' questions, and skepticism that they are informative about real-life acquisition. Most notably, large language models have attained remarkable success at discovering grammar without using any of the methods that some in linguistics insisted were necessary for a science of language to progress."

ChatGPT: To list and count words used more than once in the provided text, I'll go through the text and tally the occurrences of each word, disregarding case and punctuation.
Here is the processed count of words used more than once:
language: 7 times
models: 4 times
large: 3 times
claims: 3 times
of: 9 times
and: 5 times
the: 8 times
linguistics: 3 times
including: 3 times

In this counting, there are a number of mistakes, and I asked ChatGPT to recount the results for 'the,' 'models' and 'claims', in a separate session for each item.[7]

For the recounting of 'the,' I tried to confuse the chatbot by suggesting that 'the' is used seven times. ChatGPT replied correctly that 'the' is repeated six times, see the count of 'the' in (4).

Each time ChatGPT could understand the prompt and replied to it appropriately, but the word counts were not always correct. When I insisted on an explanation of why it makes mistakes in word counting, ChatGPT replied the following.

(3) Recounting of repeated words:

ChatGPT: I apologize for the earlier mistakes in counting words. Let's approach this systematically using a Python script to ensure we get the accurate counts for words used more than once.
Here's the Python code to count word occurrences:
text = """The rise and success of large language models undermines virtually every strong claim for the innateness of language that has been proposed by generative linguistics. Modern machine learning has subverted and bypassed the entire theoretical framework of Chomsky's

[7] The whole conversation is available at: https://chatgpt.com/share/57abf75d-5a05-4686-b603-741d1da36b5c

```
approach, including its core claims to particular insights, principles, structures, and processes. I describe the sense in which modern language models implement genuine theories of language, including representations of syntactic and semantic structure. I highlight the relationship between contemporary models and prior approaches in linguistics, namely those based on gradient computations and memorized constructions. I also respond to several critiques of large language models, including claims that they can't answer ``why'' questions, and skepticism that they are informative about real-life acquisition. Most notably, large language models have attained remarkable success at discovering grammar without using any of the methods that some in linguistics insisted were necessary for a science of language to progress."""

# Removing punctuation and converting to lowercase
import re
from collections import Counter
words = re.findall(r'\b\w+\b', text.lower())
word_counts = Counter(words)
# Filter words that appear more than once
repeated_words = {word: count for word, count in word_counts.items()
if count > 1}
repeated_words
```

Explanation of the code:

# indicates a comment.

r'\b\w+\b' is a regular expression used to match words in a string.

r' denotes a raw string, which means that the backslashes (`\`) are treated as literal characters and not as escape characters.

\b denotes a word boundary, a position where a word character is not followed/preceded by another word character.

\w matches any word character.

+ means "one or more" preceding element, i.e., **\w+** matches sequences of one or more word characters.

(4) Running the Python script from (3):

ChatGPT: Let's run this script to get the accurate counts.

Here's the accurate count of words used more than once in the provided text:

the: 6 times
and: 7 times
success: 2 times
of: 8 times
large: 3 times
language: 7 times
models: 5 times
for: 2 times
that: 4 times
has: 2 times
linguistics: 3 times
modern: 2 times
including: 3 times
claims: 2 times
to: 3 times
i: 3 times
in: 3 times
they: 2 times

Now, the number of repetitions doubled, from 9 in (2) to 18 in (4).

In sum, linguists have evaluated ChatGPT's linguistic competence with complex semantic and syntactic tasks and, as a rule, ChatGPT performs well on such tasks, see, e.g., Haider (2023). To the best of my knowledge, no linguist has challenged ChatGPT with a basic linguistic task such as word counting. Since ChatGPT does not process language in terms of words, it is miserable at word counting. To count words properly, ChatGPT needs additional software (the Python script in (3)). In another session on word counting, ChatGPT explained to me that if I wanted to avoid mistakes in word counting, my prompt should have required first a tokenization of the text, so that ChatGPT could compose words from the tokens and with the help of the free spaces preceding some of the tokens, it would then count words correctly; recall the discussion about the difference between 'er' (Figures 6.1, 6.2), and ' er' (Figures 6.4, 6.5), namely that ' er' indicates a word boundary (word beginning), marked by '\b' in the Python script in (3).

After the difficult solution of the word counting task, it could seem that ChatGPT handles language in a very complex way, more complex than what is usually done in a linguistic analysis. However, this is a misleading impression, and the next section explains why.

3. Complexity

In science, a problem often allows for different solutions. The so-called *Big O* notation serves for the assessment of the complexity of those solutions in mathematics and computer science (CS).[8]

The *Big O* notation tells us how an algorithm slows as data grows. That is, complexity is not a property of data (which is the case in linguistics, but of the algorithm [analysis]). As an illustration of this logic, let me evaluate two solutions to a task. However, it has to be mentioned here that the examples below are to help linguists understand the concept of complexity, and this explanation of the *Big O* notation is an oversimplification. In computer science, the *Big O* notation evaluates the complexity of functions and data structures (e.g., *tree* search is more complex than *list* search, cf. hierarchical syntactic trees (in LT) and linear sequences of tokens (in ChatGPT)).

(5) *Problem*: Calculate the sum of the numbers from 1 to 100.

(6) *Solution 1:* 1+2+3, and so on to 100, i.e., 99 summations are necessary to calculate the sum.

Let us check the behavior of this solution as the data grows, e.g., let us increase the amount of the data from 100 to 1000. Following the same idea (Solution 1), to calculate the sum of the numbers from 1 to 1000, we have to perform 999 summations (algorithm steps). That is, with the growth of the data, more effort is required to come to a solution.

(7) *Solution 2:* Based on the observation made by the young Gauss that 100+1 = 99+2 = 98+3, and so on to 51+50.

We can calculate the sum of the numbers from 1 to 100 only in two steps: the first step involves addition, the second step consists of multiplication: (1+100)*50=5050. An increase of the amount of the data from 100 to 1000 does not reflect on the complexity of the solution, and we can still calculate the sum of the numbers from 1 to 1000 in two steps: (1+1000)*50= 500500.

---

[8] A tutorial on the *Big O* at: https://www.geeksforgeeks.org/analysis-algorithms-big-o-analysis/

Both Solution 1 and Solution 2 give the same result, but the first solution is complex and therefore uninteresting, while Gauss's solution is simple and elegant and has been used as a formula for the sum of an arithmetic progression ever since.

How does all this relate to ChatGPT and research in linguistics? The ChatGPT approach to language relies on surface forms (orthography), Rule 1 in (8), while a linguistics approach usually relies on semantics, Rule 2 in (9).

(8) Rule 1, form-based: If a word A (e.g., Bulgarian *nivo* 'niveau') ends in -o, attach the suffix B

(e.g., -*to* in Bulgarian) to it.

(9) Rule 2, semantics-based: If X is a particular type of a verb (e.g., an action verb), derive a particular type of a noun Y (e.g., an agent) by the attachment of the productive suffix Z (e.g., -*er*).

Now, the information on which Rule 1, (8), relies is not language-specific and is directly available: the word A either terminates in -*o* or not (*nivo* ends in -*o*), and the attachment of the suffix B gives *nivoto* 'the niveau'. The semantic information on which Rule 2 (9) relies requires additional effort in order to be discovered, and Rule 2 is also language-specific, in the sense that we need some knowledge of the language from which the data come in order to apply this rule. Then, Rule 1 consists of two steps: i) we have to check whether A ends in -*o* and if yes, ii) to attach the suffix B (-*to*). Rule 2 involves the following steps: a) evaluation whether the word from the initial set is a verb; if yes, b) we have to ensure that the verb is of the type we need (an action verb), afterwards c) addition of the productive suffix -*er* to derive an agent noun, if d) the derivation is possible, because e.g. *to edit* is an action verb but it does not co-occur with -*er* (moreover, according to LT *to edit* is a backformation from *editor*, Manova, 2011a). Therefore, we conclude that Rule 2 (9) is more complex than Rule 1 (8).

Before moving to Section 4, in which I demonstrate a form-driven analysis of word-formation, let me illustrate a form-driven inflection (like in (8)) with more data from my mother tongue, Bulgarian (Slavic), (10) and (11), see also Manova (2023a). Bulgarian has a suffixal definite article, and indefinite nouns and adjectives in this language may end in -*a*. If semantics is considered, there should be four different -*a* morphemes, cf. the morphosyntactic feature values in (10) and (11), where all -*a* morphemes are bolded and indexed for convenience. The four different -*a* morphemes all select the definite article -*ta* (Manova and Dressler, 2001), though the article has allomorphs, see (1d) and the example in (8).

(10)	Nouns: indefinite	→ definite
	sg.fem: *bluz-**a**$_1$* 'blouse'	→ *bluz-**a**$_1$-**ta*** 'the blouse'
	sg.masc: *bašt-**a**$_2$* 'father'	→ *bašt-**a**$_2$-**ta*** 'the father'
	pl.neut: *sel-**a**$_3$* 'villages'	→ *sel-**a**$_3$-**ta*** 'the villages'
	cf. sg.neut: *sel-o* 'village'	→ *sel-o-to* 'the village'
(11)	Adjectives: indefinite	→ definite
	sg.fem: *krasiv-**a**$_4$* 'beautiful'	→ *krasiv-**a**$_4$-**ta*** 'the beautiful'

4. A form-based analysis of word-formation

As mentioned, (10) and (11) are inflectional morphology, and it is well-known from LT that inflection produces word forms and therefore may be accounted for without reference to meaning. By contrast, word-formation, since deriving new words, has always been seen as a semantic operation. In other words, if derivation without reference to meaning is less complex than derivation with reference to meaning, it should also be possible to account for word-formation only in terms of relations of forms, i.e., to provide an analysis of word-formation similar to how ChatGPT generates language. In this section, I demonstrate that such an analysis is indeed possible and that, actually, this is how native speakers combine word-formation suffixes.

Undoubtedly, English is the language with the most profoundly studied word-formation. (Overviews of research on word-formation from a cross-linguistic perspective in Lieber and Štekauer 2014; Plag and Balling 2016; Lieber 2017). While more recent studies analyze English word-formation based primarily, if not exclusively, on semantics (Lieber 2004, among many others), previous research known as the *Stratal approach* (Siegel 1974; Selkirk 1982; Kiparsky 1982) is form-focused, see (12): based on morphological and phonological information (the types of morpheme junction marked by '+' and '#' respectively) forms of affixes are distributed into different strata (classes) so that class II affixes are always outside class I affixes in the word-form.

(12) English: Stratal approach, from Spencer (1991, 79)

a. Class I suffixes: *+ion, +ity, +y, +al, +ic, +ate, +ous, +ive, +able, +ize*

b. Class I prefixes: *re+, con+, de+, sub+, pre+, in+, en+, be+*

c. Class II suffixes: *#ness, #less, #hood, #ful, #ly, #y, #like, #ist, #able, #ize*

d. Class II prefixes: *re#, sub#, un#, non#, de#, semi#, anti#*

Another example of a form-focused analysis is Fabb (1988). This study distributes the English suffixes into four groups as shown in (13):

(13) English: Suffix-driven selectional restrictions (Fabb 1988)

a. Group 1: suffixes that do not attach to already suffixed words

b. Group 2: suffixes that attach outside one other suffix

c. Group 3: suffixes that attach freely

d. Group 4: problematic suffixes

An alternative form-focused analysis recognizes closing suffixes: a particular suffixal form cannot be followed by other suffixes in a language (see Szymanek 2000 for English and Polish; see also Aronoff and Fuhrhop 2002). Closing suffixes have been established in a number of languages. Manova (2015b) is an overview of research on the topic.

A form-based restriction on the order of English derivational suffixes of a different kind is reported in Manova (2011b) and Manova and Knell (2021). Manova (2011b) sees derivational suffix combinations as binary structures of the type SUFF1-SUFF2, where SUFF1 has three valency positions for further suffixation: $\text{SUFF2}_{\text{Noun}}$, $\text{SUFF2}_{\text{Adjective}}$ and $\text{SUFF2}_{\text{Verb}}$. The idea of this distribution of outputs according to the lexical-category specification of SUFF2 is based on the Gauss-Jordan elimination, a mathematical method for solving systems of linear equations numerically, that is, only with the help of elementary operations such as substitution, addition or multiplication. (14) is an example of a system of linear equations.

(14) $2x + y + 2z = 10$

$x + 2y + z = 8$

$3x + y - z = 2$

The goal of Gauss-Jordan is, based only on well-known facts and elementary operations with them, to come to a single option for a variable (the unknown); *x*, *y* and *z* are the variables in (14). If there is only one option for a variable, this option is the solution to the problem.

With respect to affix order in word-formation, the well-known information is information about the lexical category specification of an affix, i.e., whether the affix derives nouns (N), adjectives (A) or verbs (V); a single option for a variable means one affix combination of a kind. As can be seen from Table 6.1, this method allows linguistic data to be distributed so that, in most cases, there is indeed one option of a kind: for N (*-ist*$_N$*-dom*$_N$) and for V (*-ist*$_N$*-ize*$_V$). I label such combinations *fixed.*

Table 6.1: Combinability of the English suffix *-ist*

SUFF1	Lexical category of SUFF1	SUFF2 classified for lexical category. in brackets, number of types (lemmas) derived with the combination SUFF1-SUFF2	
-ist	N	N: *-dom* (2)	[*fixed combination*]
		A: *-ic* (631), *-y* (5)	[*predictable combination*]
		V: *-ize* (3)	[*fixed combination*]

Source: Previous research of mine; data from Aronoff and Fuhrhop, 2002, based on OED, CD, 1994.

If more than one SUFF2 of the same lexical category is available (see for A(djective) in Table 6.1), one of the SUFF2 suffixes attaches by default, suffix -*ic*$_A$ in our case: in English, the combination *-ist*$_N$*-ic*$_A$ derives 631 types, while -*ist*$_N$*-y*$_A$ derives only 5 types, which makes *-ist*$_N$*-ic*$_A$ a *predictable* combination.[9] Regarding default suffixes, having counted suffix combinations in large dictionaries and corpora for different languages, Manova (2011b, 2015a), Manova and Talamo (2015), and Manova and Knell (2021) maintain that a default suffix derives more than ten types (such suffixes are referred to as productive), while SUFF2 suffixes that compete with the default suffix derive ten or fewer types each. Thus, in Table 6.1, different forms of suffix combinations correspond to different meanings (by default), which means a one-to-one mapping of meaning and form and allows us to identify the suffix combinations at hand both in terms of form (cf. the ChatGPT approach) and in terms of meaning (the LT approach).

Table 6.2 applies the logic of Gauss-Jordan to a more complex case, the combinability of the Polish suffix *-arz.*[10] Polish, unlike English, is an inflecting fusional language, and derivational suffixes are often followed by inflection, i.e., in Polish, the inflection is obligatory for the well-formedness of a word. All

---

[9] According to ChatGPT, both 'istic' and 'isty' are single tokens, [4633] and [88623], respectively, which confirms my reasoning.

[10] I thank Bartosz Brzoza for his help with the Polish data.

inflectional suffixes in Table 6.2 are in brackets. Descriptions and analyses of Polish derivational morphology by Polish scholars, as a rule, give derivational suffixes together with the inflection that follows them, either in brackets, as done in Table 6.2, or unmarked, as a single suffix with the derivational one, e.g., as *-n(y)* or *-ny*, respectively; see the first adjectivizing SUFF2 in Table 6.2. (For a semantics-based analysis of the combinability of Polish word-formational suffixes, see Burkacka 2015; see also the discussion of Polish word-formation in Szymanek 2010.)

As shown in Table 6.2, the suffix *-arz* combines with more than one adjectivizing SUFF2 and with a set of nominalizing SUFF2 suffixes. While for the derivation of adjectives, there is only one default suffix, *-sk(i)* (>10), three different nominalizing suffixes that derive more than ten types can follow the suffix *-arz*: *-czyk* (>10), *-ni(a)* (>10) and *-stw(o)* (>10), all bolded in Table 6.2 for convenience. The existence of three default suffixes of the same type (nominalizers), all "competing" for *-arz* seems to challenge my analysis. However, the three suffixes differ in both form and meaning: *-czyk* (>10), default for derivation of *persons*; *-ni(a)* (>10), default for derivation of *places*; and *-stw(o)* (>10), default for *abstract/collective nouns*. That is, neither homophony nor polysemy is involved in the suffix combinations in Table 6.2. I therefore conclude that all suffix combinations in Table 6.2 are predictable, and since they are instances of one-to-one mapping of form and meaning (by default), they can also be identified only in terms of form (the ChatGPT approach).

Considering the fact that derivational suffixes in English and Polish seem to form only fixed and predictable combinations, i.e., should be identifiable in terms of form, I hypothesized that native speakers should have memorized those combinations and, consequently, should be able to produce them without reference to meaning, that is, based exclusively on form. To test this hypothesis, I designed a psycholinguistic experiment, the results of which are reported below. Here I present only the results of the native speakers of English and Polish, but the experiment was also conducted with native speakers of German, Italian, Spanish and Slovene, and with advanced non-native speakers of English and German. Overall, the results of all iterations converge. (For curious readers, the scores of the non-native speakers of English are reported in Manova and Knell 2021; the scores of the native and non-native speakers of German can be found in Brosche and Manova 2022).

Table 6.2: Combinability of the Polish suffix *-arz*

SUFF1	Lexical category of SUFF1	Lexical category of SUFF2	SUFF1-SUFF2 exemplified	Notes
-arz	N	i. ADJ: *-n(y)* (2)	moc-*ar-n(y)* 'strong'	[derives only 2 adjectives]
		ii. ADJ: *-ow(y)* (1)	gęśl-*arz-ow(y)* 'of fiddler'	[derives a single adjective]
		iii. **ADJ: *-sk(i)* (>10)**	pis-*ar-sk(i)* 'of writer'	[**default** for derivation of adjectives]
		a. N**: *-czyk* (>10)**	piek-*ar-czyk* 'baker's apprentice'	[**default** for derivation of **persons**, cf. f]
		b. N: *-k(a)* (2)	mur-*ar-k(a)* 'bricklaying'	[derives only 2 abstract nouns, cf. e]
		c. **N: *-ni(a)* (>10)**	kreśl-*ar-ni(a)* 'drafting studio'	[derives nouns for **places**]
		d. N: *-nik* (1)	piek-*ar-nik* 'oven'	[derives a single **object**]
		e. **N: *-stw(o)* (>10)**	księg-*ar-stw(o)* 'all booksellers'	[**default abstract/collective nouns**, cf. b]
		f. N: *-yn(a)* (5)	mur-*arz-yn(a)* 'bad bricklayer'	[derives only 5 nouns for persons, cf. a]

4.1 Method

64 native Polish speakers and 45 native English speakers were tested; they all participated on a voluntary basis. The questionnaire presented to them consisted of three parts:

- A series of general demographic questions regarding age, gender, nationality, native language(s), other languages spoken, level of education, and experience in a linguistic or other language-related field.
- A small practice to ensure that the participants understood the task properly. The training examples were not part of the test stimuli.
- The main task: 60 suffix combinations (e.g., *-istic* in English, *-arny* in Polish) were presented in a randomized order, and participants were asked to identify which of the combinations exist and which do not exist as word terminations in the respective language. Of the 60 combinations,

30 existed in the respective language, and 30 did not. Of the existing combinations, 15 were productive (derived more than ten types), and 15 were unproductive. Of the non-existing combinations, 15 were created by a permutation of an existing combination (reversing the order of the two suffixes such that the combination was not possible in the respective language), and 15 were created through a spelling manipulation of an existing combination (changing one letter from an existing combination such that the new form does not exist in the particular language). No non-existing combinations included any phonological and/or orthographical impossibilities in the respective language.

4.2 Data analysis

Independent t-tests were used to consider possible significance of overall scores, as well as for stimulus type: existing vs. non-existing and productive vs. unproductive combinations. Figure 6.14 presents the results of the native speakers of English and Polish.

Figure 6.14: Native speakers' accuracy of recognition of the 60 suffix combinations tested in the experiment (only statistically significant results).

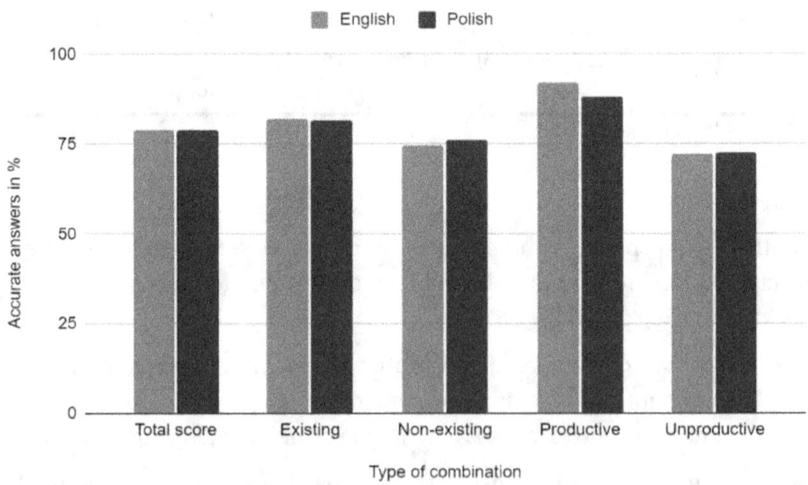

Total score = correct answers for all types of suffix combinations tested: existing combinations of two types (productive and unproductive) and non-existing of two types (permutations and manipulations, see Method).

The participants in the experiment did not need semantic cues to process suffix combinability, i.e., they could differentiate between existing and non-existing suffix combinations presented to them without lexical bases such as roots/stems/words. Statistically significant were the differences between existing and non-existing combinations, and between productive and unproductive combinations. As already mentioned, English and Polish differ typologically in the sense that English has very poor inflectional morphology, while Polish is characterized by a very rich inflectional system. Nevertheless, the results obtained for the two languages are virtually the same, the total score of the correct answers for English is 79% and 78.86% for Polish (Figure 6.14), though combinations of three suffixes (trigrams, the case of Polish where two derivational suffixes are often followed by inflection) should be easier to recognize than combinations of two suffixes (bigrams, the case of English derivational suffix combinations). In other words, inflection did not seem to have an impact on the processing of suffix combinability in word-formation. I therefore conclude that native speakers of Polish see inflection as forming a natural subword unit with the derivational material that precedes it; on this issue, see also Manova (2023b).

Since suffix combinability is not taught at school and all linguistic theories assume that a morphological derivation always starts with a root/stem, depending on the theory, the only plausible explanation why native speakers of English and Polish successfully accomplished a task they should not be able to solve is that they had extracted and subconsciously memorized adjacent suffixes in terms of bigrams[11] and trigrams, during language acquisition (cf. tokenization in ChatGPT).

5. Discussion

In this section, I address the question of whether ChatGPT can serve as a theory of language. I claim that the answer to this question is "yes," especially if the new theory is allowed to mix insights from different approaches to morphology. I discuss the morpheme-based Distributed Morphology (Halle and Marantz 1993, Harley and Noyer 1999, Embick and Noyer 2007, Bobaljik 2017) and the paradigm-based Paradigm Function Morphology (Stump 2001, 2016, Stump and Finkel 2013, Bonami and Stump 2017). The former denies the importance of the paradigm as an analytical tool; the latter does the same with the

---

[11] Analyses of affix order in terms of bigrams are proposed in Ryan (2010) and Mansfield et al. (2020).

morpheme. Additionally, DM is syntax-centered, i.e., syntactic rules derive morphological structure, while PFM claims independence of morphology from syntax. I therefore consider DM and PFM a good example of competing theories. Of course, the two theories also have things in common, e.g., they both are meaning-first. A DM derivation concatenates abstract morphemes that have meaning but no form. Such morphemes are placed in terminal nodes on the syntactic tree, and the association with phonological form is postponed, the so-called *late insertion.* PFM, in its turn, assumes the existence of content paradigms (ordered paradigm cells identifiable in terms of sets of morphosyntactic features); the content paradigm is then linked to a corresponding form paradigm. (In some versions of the theory, the existence of a third paradigm called realizational is assumed; the form-paradigm copes with meaning-form mismatches, the realizational paradigm gives the final forms.)

Before starting with the theoretical discussion, let me mention some a-theoretical issues typical of language in general. ChatGPT, similar to human languages, has a set of letters and symbols that serve for vocabulary initialization, i.e., for the production of larger elements, tokens or subword units (recall the List of 100256 tokens). A human language in a written form is organized in the same way, in the sense that it also possesses letters (an alphabet and also uses some additional symbols for numbers, etc.) and derives basic units (morphemes) as combinations of letters. Note that morphemes are, like tokens, subword units. Thus, with respect to letters and basic units built from then, ChatGPT and human languages appear to be very similar, if not the same.

If we think of ChatGPT in the context of theories of language, a highly relevant issue is how the model learns vocabulary items. In ChatGPT, the process is entirely frequency-governed (cf. BPE): the frequency of occurrence of adjacent letters determines whether the letter combination they form will be added to the vocabulary or not. Can such a scenario serve for language acquisition (LA) by humans? It seems to me that the answer to this question should be "yes," though, to the best of my knowledge, there is no research on the topic. Additionally, linguists should be aware that if LA is researched only in terms of morphemes and words, as done in LT, this cannot lead to the conclusion that LA is, like ChatGPT, token-based.

Now on theoretical issues, the first significant difference between ChatGPT and LT consists in the fact that ChatGPT's basic units, the tokens, are not associated with semantics, which thus makes ChatGPT incompatible with DM but in harmony with PFM where word-internal structure consists of markings that have only form and no meaning (recall that in PFM meaning is assigned only to paradigm cells).

Regarding the identification of the basic units, ChatGPT tokens may be morphemes and words, though letters that serve for the production of tokens are also listed in the vocabulary and referred to as 'tokens.' This fact is compatible with DM because in this theory, the same syntactic rules are responsible for the combinations of morphemes and words, which thus makes morphemes and words the same type of unit. PFM, since rejecting morphemes and assigning semantics only to whole words occupying paradigmatic cells, is incompatible with ChatGPT on this criterion.

As for the demarcation of derivation and inflection, similar to DM, ChatGPT, does not make a distinction between derivation and inflection and tokens may be non-derived words, derived words or inflected words, e.g., *read, reads, organization* and *organizations* are four single tokens, Figures 6.15 and 6.16 for ' read' and ' reads,' respectively, and Figures 6.17 and 6.18 for ' organization' and ' organizations.' In other words, all four words are frequent enough to be part of the vocabulary and therefore count as single tokens.

Figure 6.15: 'read' is a single token

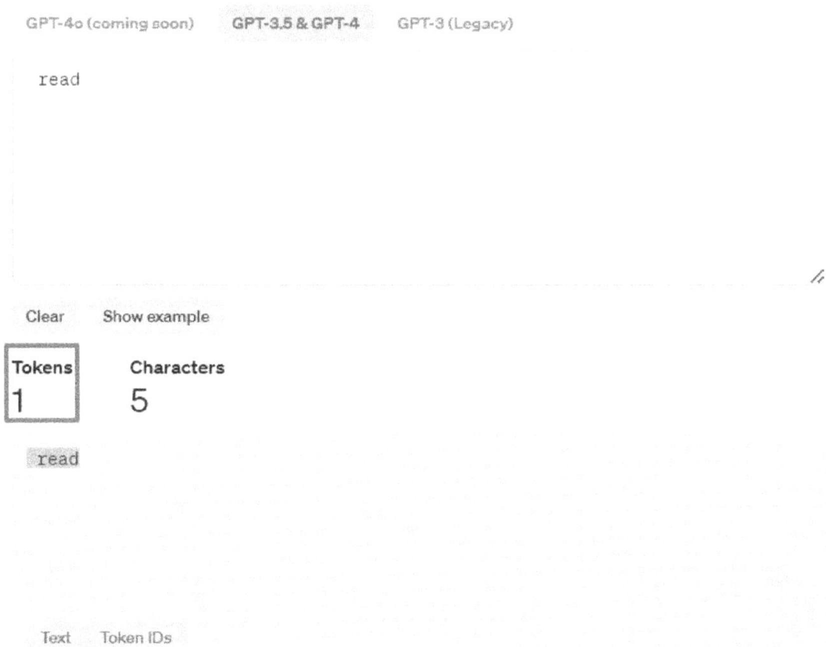

Figure 6.16: 'reads' is a single token

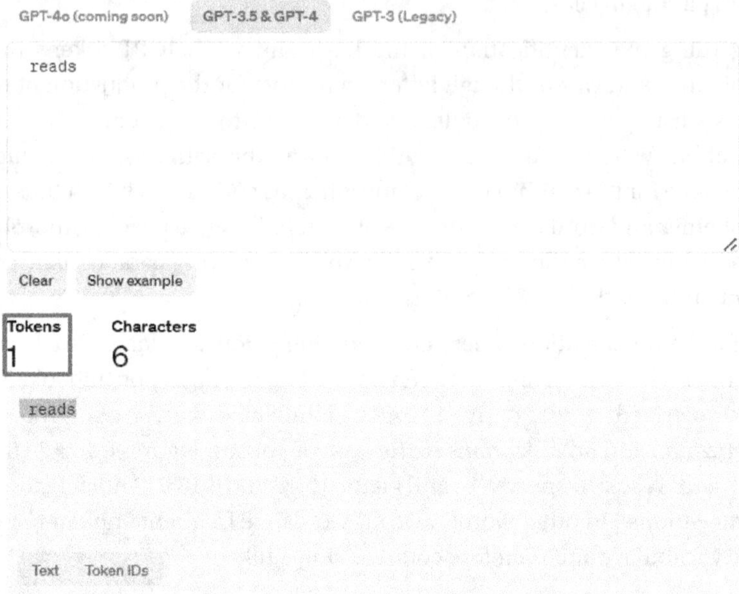

Figure 6.17: 'organization' is a single token

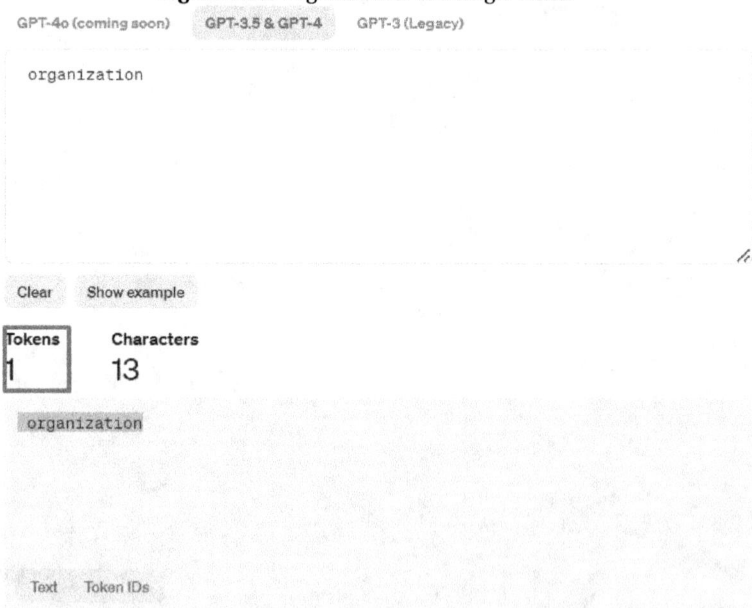

Figure 6.18: 'organizations' is a single token

There are also facts about ChatGPT that are incompatible with both theories of morphology discussed, DM and PFM. For example, ChatGPT's basic units (tokens) are a fixed-size set but emerge from data (are extracted from a large amount of text based on the frequency of adjacent letters without reference to meaning (recall the logic of the BPE algorithm), while morphemes in LT are more or less a predefined set too but one can enlarge this set with, e.g., morphemes borrowed from other languages, or with newly created morphemes, and the enrichment of the initial set of morphemes does not have anything to do with frequency of use of adjacent characters, which is the only criterion for vocabulary inclusion in ChatGPT. Moreover, based on the inventory of ChatGPT's basic units, we can conclude that ChatGPT sees phonology, morphology and syntax as taking place simultaneously; while in LT, morphology either precedes (the case of PFM) or follows syntax (the case of DM, where actually both morphology and phonology are postsyntactic). Then, as demonstrated in Section 2, tokens do not always coincide with linguistic units. To understand the issue, let us scrutinize the tokenization of *illegible, illogical, illegal,* and *illiberal,* Figures 6.19 to 6.22. According to LT, all four words are derived with the prefix *il-*. ChatGPT, however, parses the four words in four different ways and acknowledges a prefix ' il' / *il-* only in *illogical.* Nevertheless, ChatGPT's parsing of the four *il-* words is compatible with the results of linguistic research in parsability, the so-called *Parsability Hypothesis* (Hay 2002). Hay claims the following:

> "If the derived form is more frequent than the base it contains (*illegible* is more frequent than *legible*), then the whole-word route will have an advantage. If the derived form is less frequent than the base it contains (*illiberal* is less frequent than *liberal*), then the decomposed route will be advantaged." (Hay 2002, 529)

ChatGPT's tokenization appears to have the same logic, just Hay overlooked that *illegible* does not have only a prefix but also a very frequent suffix, thus *il-* is not parsed (as Hay predicts) but *-igible* is, which is actually in accord with the Parsability Hypothesis. In the Tokenizer, it is *illeg-ible,* Figure 6.19. As regards *illiberal,* it is indeed decomposed as Hay predicts, but the parsing in the Tokenizer differs from Hay's: *ill-ib-eral,* Figure 6.22. The tokenization of *ill-ib-eral* is, of course, due to the high frequency of its constituents, which is entirely compatible with Hay's research; only Hay assumes that English has a predefined set of morphemes, and this does not allow her to acknowledge subword units that are highly frequent but not in the set.

Figure 6.19: 'illegible'

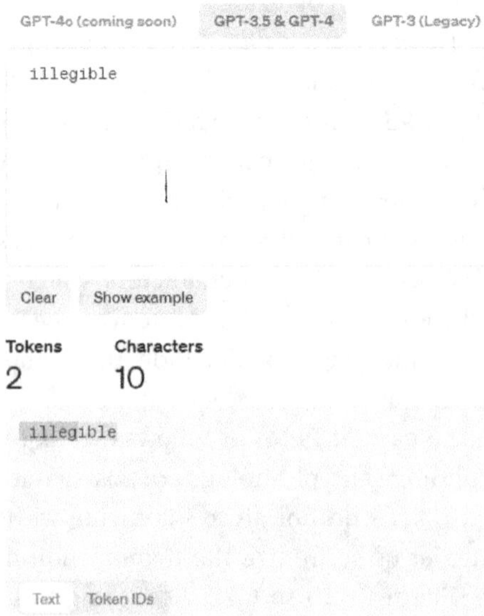

Figure 6.20: 'illogical'

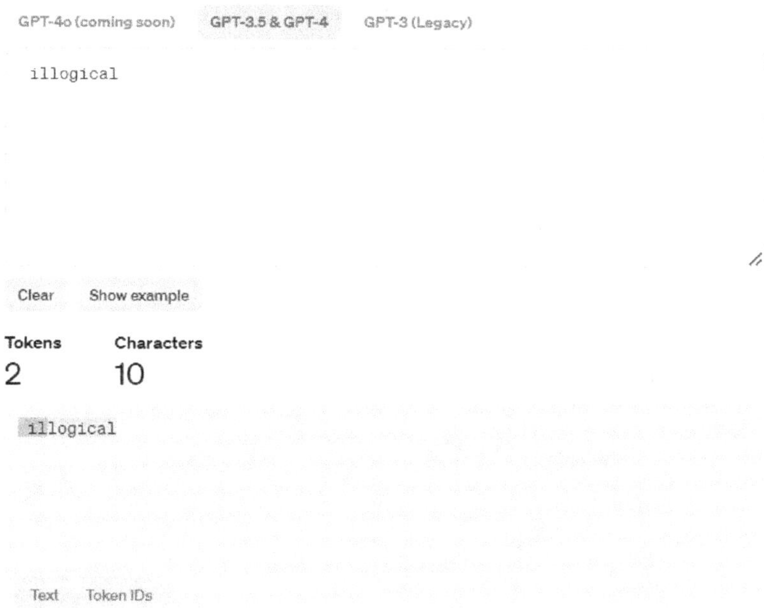

Figure 6.21: 'illiberal'

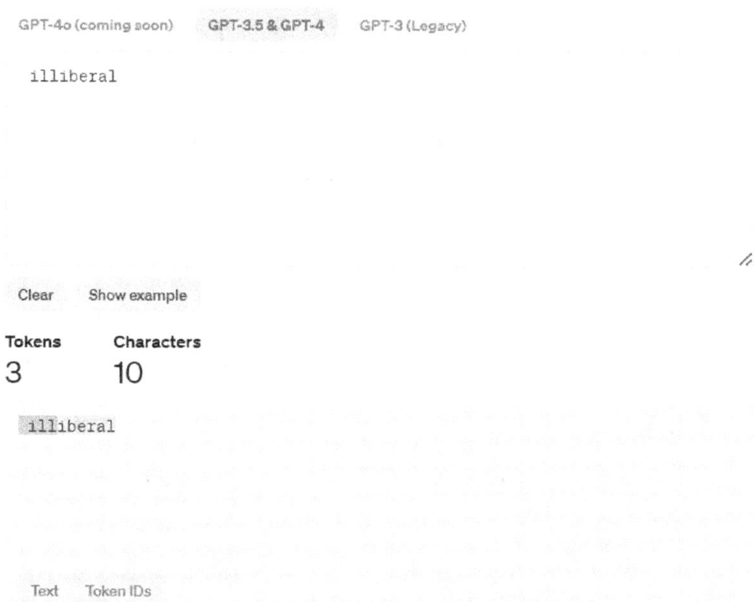

Figure 6.22: 'illegal'

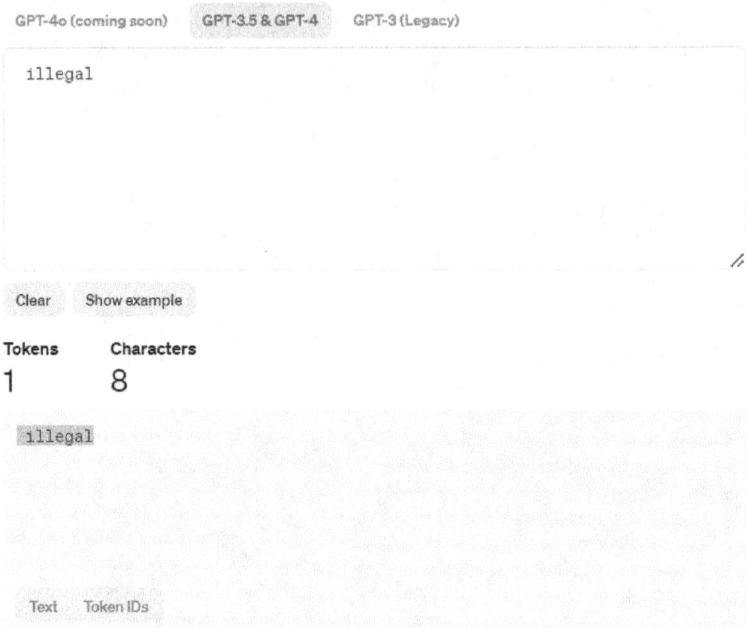

So we come to the length of forms used by ChatGPT and in linguistic morphology. As a rule, ChatGPT's prompts are long and consist of more than one token, i.e., prompts are (comparable with) sentences. By contrast, in morphology, the longest sequences of basic units (morphemes) are words. Significantly, ChatGPT does not encounter form-meaning mismatches, while morphology in LT is about form-meaning mismatches. Thus, if ChatGPT can generate language based only on form, form and meaning in language are in a perfect relationship, i.e., form-meaning mismatches that are the central issue in morphological research seem to be fictitious problems that arise from data presentation (the small length of sequences analyzed). Moreover, in LT, it is well-known that disambiguation of morphemes (15) and words (16) requires longer sequences of units of the same kind.

(15) Disambiguation of *-er*

read → *read-er* → *read-er-ship* (nominal *-er*)

versus

large → *larg-er* → **larg-er-ship* (adjectival *-er*)

(16) Disambiguation of *read*

I read (*read* is a verb)

versus

I gave a read of (*read* is a noun)

In sum, the form-meaning mismatches that are the central topic of research in morphology are due to the small sequences of form analyzed.

Finally, ChatGPT does not rely on abstract morphemes and syntactic trees (the case of DM);[12] it does not recognize abstract lexemes and morphological paradigms, either (the case of PFM)—because working with abstract units such as lexemes and complex structures such as paradigms and trees is always more complex than working with real, existing forms organized in linear sequences (recall the discussion on complexity in Section 3). Note also that ChatGPT's long sequences of tokens have their parallel in LT: verbs in polysynthetic languages (Fortescue et al. 2017), which are similar to sentences in English.

For the sake of completeness and fairness, Piantadosi (2023), with the title of his paper "Modern language models refute Chomsky's approach to language," put the focus on the discussion on the relationship between LLMs and LT on Chomsky's approach. However, LLMs reveal problems with other theoretical frameworks as well. For example, the paradigmatic approach is also challenged because, from a ChatGPT perspective, using paradigms for language processing is inefficient, in the sense that if language is generated step-by-step in terms of a linear sequence of forms, and we look for the next piece in that sequence, we need only a single piece of form, not a paradigm. Then, the forms in a paradigm are, as a rule, not substitutable in a sentence slot, i.e., they do not compete with each other during sentence generation. Therefore, it does not make sense to activate the whole paradigm in order to find the single form needed. With the inclusion of inflected words in its vocabulary (Figures 6.15 to 6.18), ChatGPT provides a solution to the problem: no search for a specific form in a paradigm but a direct access to this form in the vocabulary (the lexicon in LT terminology).

The goal of this section is not to provide an exhaustive list of differences and commonalities between ChatGPT and LT but to demonstrate a way of thinking that makes possible the use of ChatGPT as a theory of language. I hope to have demonstrated that ChatGPT is compatible with LT.

---

[12] On the problems with syntactic trees (they, among other things, grow in the wrong direction, from leaves to the root), see Manova (2022).

6. Conclusion

The basic units with which ChatGPT understands and generates language are the tokens (subword units). ChatGPT's vocabulary contains 100k tokens. I started with the introduction of the Byte Pair Encoding algorithm, which serves for tokenization in ChatGPT. I showed that tokens can coincide in form with letters, morphemes or words, as well as with non-linguistic units. Consequently, tokens are not associated with semantics. Following the ChatGPT's approach of form-based language processing and my previous research in morphology, I demonstrated a form-driven analysis of word-formation and supported it with the results of a psycholinguistic experiment with native speakers of English and Polish: similar to ChatGPT, native speakers process combinations of word-formation suffixes without reference to meaning. I used the technical details of ChatGPT's internal organization to compare the model with two major theories of morphology, the morpheme-based DM and the paradigm-based PFM. ChatGPT appears to mix insights from both theories. The model not only can serve as a theory of language but can also provide a novel perspective on existing theories and reveal shortcomings. With respect to morphology, the logic behind ChatGPT's language processing shows that form-meaning mismatches, the central topic of research in morphology, are fictitious problems, in the sense that they are due to the way data are presented. The use of long(er) sequences of form, like in ChatGPT prompts, effortlessly resolves form-meaning mismatches.

Bibliography

Aronoff, Mark, and Nanna Fuhrhop. 2002. "Restricting Suffix Combinations in German and English: Closing Suffixes and the Monosuffix Constraint." *Natural Language and Linguistic Theory* 20: 451–90. https://doi.org/10.1023/A:1015858920912

Bobaljik, Jonathan D. 2017. "Distributed Morphology." *Oxford Research Encyclopedia in Linguistics.* Oxford University Press. https://doi.org/10.1093/acrefore/9780199384655.013.131

Bonami, Olivier, and Gregory Stump. 2017. "Paradigm Function Morphology." In *The Cambridge Handbook of Morphology*, edited by Andrew Hippisley and Greg Stump, 449–81. Cambridge: Cambridge University Press. https://doi.org/10.1017/9781139814720.017

Brosche, Kimberly, and Stela Manova. 2022. "German Word Formation and the Organization of the Mental Lexicon." Paper presented at the Annual Conference of the Middle European Master Program in Cognitive Science, Zagreb, June. https://homepage.univie.ac.at/stela.manova/uploads/1/2/2/4/12243901/poster_suffix_combinations_version.pdf

Burkacka, Iwona. 2015. "Suffix Sets in Polish De-Nominal Derivatives." In *Affix Ordering Across Languages and Frameworks*, edited by Stela Manova, 233–58. Oxford: Oxford University Press. https://doi.org/10.1093/acprof:oso/9780190210434.003.0010

Chomsky, Noam, Ian Roberts, and Jeffrey Watumull. 2023. "The False Promise of ChatGPT." *New York Times*, March 8, 2023. https://www.nytimes.com/2023/03/08/opinion/noam-chomsky-chatgpt-ai.html

Embick, David, and Rolf Noyer. 2007. "Distributed Morphology and the Syntax/Morphology Interface." In *The Oxford Handbook of Linguistic Interfaces*, edited by Gillian Ramchand and Charles Reiss, 289–324. Oxford: Oxford University Press. https://doi.org/10.1093/oxfordhb/9780199247455.013.0010

Fabb, Nigel. 1988. "English Suffixation Is Constrained Only by Selectional Restrictions." *Natural Language and Linguistic Theory* 6 (4): 527–39. https://doi.org/10.1007/BF00134494

Haider, Hubert. 2023. "Is Chat-GPT a Grammatically Competent Informant?" *lingbuzz* preprint lingbuzz/007285. https://lingbuzz.net/lingbuzz/007285

Halle, Morris, and Alec Marantz. 1993. "Distributed Morphology and the Pieces of Inflection." In *The View from Building 20: Essays in Linguistics in Honor of Sylvain Bromberger*, edited by Kenneth Hale and Samuel Jay Keyser, 111–76. Cambridge, MA: MIT Press.

Harley, Heidi, and Rolf Noyer. 1999. "State-of-the-Article: Distributed Morphology." *GLOT International* 4 (4): 3–9.

Hay, Jennifer. 2002. From speech perception to morphology: Affix-ordering revisited. *Language* 78 (3): 527–555.

Katzir, Roni. 2023. "Why Large Language Models Are Poor Theories of Human Linguistic Cognition: A Reply to Piantadosi (2023)." *Biolinguistics* 17: e13153. https://doi.org/10.5964/bioling.13153

Kiparsky, Paul. 1982. Lexical Morphology and Phonology. In *The Linguistic Society of Korea, Linguistics in the Morning Calm*, 1-91. Seoul: Hanshin Publishing.

Lieber, Rochelle. 2004. *Morphology and Lexical Semantics*. Cambridge: Cambridge University Press.

Lieber, Rochelle. 2017. "Derivational Morphology." *Oxford Research Encyclopedia of Linguistics*. Oxford University Press. https://doi.org/10.1093/acrefore/9780199384655.013.248

Lieber, Rochelle, and Pavol Štekauer, eds. 2014. *The Oxford Handbook of Derivational Morphology*. Oxford: Oxford University Press. https://doi.org/10.1093/oxfordhb/9780199641642.001.0001

Manova, Stela. 2011a. *Understanding Morphological Rules*. London: Springer. https://doi.org/10.1007/978-90-481-9547-3

Manova, Stela. 2011b. "A Cognitive Approach to SUFF1–SUFF2 Combinations: A Tribute to Carl Friedrich Gauss." *Word Structure* 4: 272–300. https://doi.org/10.3366/word.2011.0014

Manova, Stela. 2015a. "Affix Order and the Structure of the Slavic Word." In *Affix Ordering Across Languages and Frameworks*, edited by Stela Manova, 205–30. Oxford: Oxford University Press. https://doi.org/10.1093/acprof:oso/9780190210434.003.0009

Manova, Stela. 2015b. "Closing Suffixes." In *Word-Formation in the European Languages*, Vol. 2, *Handbooks of Linguistics and Communication Science (HSK) 40/2*, edited by Peter O. Müller, Ingeborg Ohnheiser, Susan Olsen, and Franz Rainer, 956–71. Berlin: De Gruyter Mouton.

Manova, Stela. 2022. "The Linear Order of Elements in Prominent Linguistic Sequences: Deriving Tns–Asp–Mood Orders and Greenberg's Universal 20 with N-Grams." *lingbuzz* preprint lingbuzz/006082. https://lingbuzz.net/lingbuzz/006082

Manova, Stela. 2023a. "Ordering Restrictions Between Affixes." In *The Wiley Blackwell Companion to Morphology*, edited by Peter Ackema, Sabrina Bendjaballah, Eulàlia Bonet, and Antonio Fábregas. Hoboken, NJ: John Wiley & Sons. https://doi.org/10.1002/9781119693604.morphcom058

Manova, Stela. 2023b. "ChatGPT, N-Grams and the Power of Subword Units: The Future of Research in Morphology." In *Proceedings of DeriMo 2023*, edited by Matea Filko and Krešimir Šojat, 1–12. Zagreb: Croatian Language Technology Society. https://lingbuzz.net/lingbuzz/007598

Manova, Stela, and Wolfgang U. Dressler. 2001. "Gender and Declensional Class in Bulgarian." *Wiener Linguistische Gazette* 67–69: 45–81. https://doi.org/10.1177/001654929004500106

Manova, Stela, Harald Hammarström, Itamar Kastner, and Yining Nie. 2020. "What Is in a Morpheme? Theoretical, Experimental and Computational Approaches to the Relation of Meaning and Form in Morphology." *Word Structure* 13 (1): 1–21. https://doi.org/10.3366/word.2020.0167

Manova, Stela, and Georgia Knell. 2021. "Two-Suffix Combinations in Native and Non-Native English: Novel Evidence for Morphomic Structures." In *All Things Morphology: Its Independence and Its Interfaces*, edited by Sedigheh Moradi, Marcia Haag, Janie Rees-Miller, and Andrija Petrović, 305–23. Amsterdam: John Benjamins. https://doi.org/10.1075/cilt.352.13man

Manova, Stela, and Luigi Talamo. 2015. "On the Significance of the Corpus Size in Affix-Order Research." *SKASE Journal of Theoretical Linguistics* 12 (4): 369–97.

Mansfield, John, Sabine Stoll, and Balthasar Bickel. 2020. "Category Clustering: A Probabilistic Bias in the Morphology of Verbal Agreement Marking." *Language* 96 (2): 255–93. https://doi.org/10.1353/lan.2020.0024

Moro, Andrea, Matteo Greco, and Stefano F. Cappa. 2023. "Large Languages, Impossible Languages and Human Brains." *Cortex* 167: 82–85. https://doi.org/10.1016/j.cortex.2023.07.003

OpenAI. 2023. "GPT-4 Technical Report." *arXiv* preprint arXiv:2303.08774. https://arxiv.org/abs/2303.08774

Piantadosi, Steven T. 2023. "Modern Language Models Refute Chomsky's Approach to Language." *LingBuzz* preprint lingbuzz/007180. https://lingbuzz.net/lingbuzz/007180

Plag, Ingo, and Laura Balling. 2016. "Derivational Morphology: An Integrative Perspective on Some Fundamental Questions." In *Word Knowledge and Word Usage: A Cross-Disciplinary Guide to the Mental Lexicon*, edited by Vito Pirrelli, Ingo Plag, and Wolfgang U. Dressler, 295–335. Berlin: De Gruyter. https://doi.org/10.1515/9783110440577-008

Rawski, Jon, and Lucie Baumont. 2023. "Modern Language Models Refute Nothing." *LingBuzz* preprint lingbuzz/007203. https://lingbuzz.net/lingbuzz/007203

Ryan, Kevin M. 2010. "Variable Affix Order: Grammar and Learning." *Language* 86 (4): 758–91. https://doi.org/10.1353/lan.2010.0026

Sauerland, Uli. 2023. "The Impact of Large Language Models on Linguistic Theory and Generative Grammar: A Critical Analysis." *LingBuzz* preprint lingbuzz/007217. https://lingbuzz.net/lingbuzz/007217

Selkirk, Elisabeth. 1982. *The Syntax of Words.* Cambridge, MA: MIT Press.

Sennrich, Rico, Barry Haddow, and Alexandra Birch. 2016. "Neural Machine Translation of Rare Words with Subword Units." *arXiv* preprint arXiv:1508.07909v5. https://arxiv.org/abs/1508.07909

Siegel, Dorothy. 1974. *Topics in English Morphology.* Cambridge, MA: MIT Press.

Spencer, Andrew. 1991. *Morphological Theory.* Oxford: Blackwell.

Stump, Gregory T. 2001. *Inflectional Morphology: A Theory of Paradigm Structure.* Cambridge: Cambridge University Press. https://doi.org/10.1017/CBO9780511486333

Stump, Gregory T. 2016. *Inflectional Paradigms: Content and Form at the Syntax–Morphology Interface.* Cambridge: Cambridge University Press. https://doi.org/10.1017/CBO9781316105290

Stump, Gregory T., and Raphael A. Finkel. 2013. *Morphological Typology: From Word to Paradigm.* Cambridge: Cambridge University Press. https://doi.org/10.1017/CBO9781139248860

Szymanek, Bogdan. 2000. "On Morphotactics: Closing Morphemes in English." In *PASE Papers in Language Studies*, edited by Bożena Rozwadowska, 311–20. Wrocław: Aksel.

Szymanek, Bogdan. 2010. *A Panorama of Polish Word-Formation.* Lublin: Wydawnictwo KUL.

Chapter 7

Evaluating the existence proof: LLMs as cognitive models of language acquisition*

Héctor Javier Vázquez Martínez

University of Pennsylvania

Annika Heuser

University of Pennsylvania

Charles Yang

University of Pennsylvania

Jordan Kodner

Stony Brook University

Abstract: The success of neural language models (LMs) on many technological tasks has brought about their potential relevance as scientific theories of language, despite some clear differences between LM training and child language acquisition. The authors argue that some of the most prominent benchmarks for evaluating the syntactic capacities of LMs may not be sufficiently rigorous. In particular, they show that the template-based benchmarks lack the structural diversity commonly found in the theoretical and psychological studies of language. When trained on small-scale data modeling of child language acquisition, the LMs can be readily matched by simple baseline models. The authors advocate for the use of the readily available, carefully curated datasets that have been evaluated for gradient acceptability by large pools of native speakers and are designed to probe the

* This work is a substantial revision of a paper presented at the 1st GenBench Workshop on (Benchmarking) Generalization collocated with the 2023 EMNLP conference held in Singapore (Vázquez Martínez et al. 2023). We thank the reviewers and the audience of the workshop for helpful feedback.

structural basis of grammar specifically. On one such dataset, the LI-Adger dataset, LMs evaluate sentences in a way inconsistent with human language users. The chapter concludes with suggestions for better connecting LMs with the empirical study of child language acquisition.

Keywords: Language Acquisition, Large Language Models, Artificial Intelligence, Natural Language Processing

***

1. The promise of existence proofs

The rapid development of deep learning and neural large language models (LLMs) over the past decade has spurred interest in their relevance to the cognitive science of language (Linzen and Baroni 2021). An assortment of performance benchmarks, many of which will be reviewed below, have been developed to evaluate the capacity of LLMs — and with apparent success. Of course, even human-like performance does not imply that LLMs are in fact appropriate cognitive models of human language and cognition. Most obviously, LLMs require orders of magnitude more input than human children receive: BERT was trained on about 3.3 billion words, and Chinchilla on 1.4 trillion, and the recent editions of GPT models likely on orders of magnitude more, while an English-learning child only receives about a few million words per year, for a total vocabulary measured in the hundreds at age three (Fenson et al. 1994; Bornstein et al. 2004).

In a welcome move, recent studies have attempted to bring LLMs closer to the psychological setting of language acquisition. Aside from enhancing connections to the scientific study of language, smaller LLMs may also contribute to the development of technologies that serve low(er)-resourced linguistic communities. A concrete example was provided by Huebner et al. (2021): a specially tuned model trained on only 5M tokens of child-directed speech (CDS) performs well on an evaluation dataset adapted from a widely used behavioral benchmark originally designed to evaluate much larger LLMs (Warstadt et al. 2020). And in 2023, an aptly-named shared task, the BabyLM Challenge, asked participants to limit training data to 100 million words (about the input of an adolescent). The message is clear: Now that huge LLMs have already succeeded, it is time to look at small ones (Warstadt et al. 2023).

For some researchers, a successful child-sized LLM promises new understandings of human language and its acquisition. According to Portelance and Jasbi (2023), for example, LLMs would provide a proof-of-concept for what is possible "in practice" for machines and "in principle" for human learners. Warstadt and Bowman (2022) go a step further: They hold that LLMs can help understand child language acquisition as long as they succeed on *some* behavioral benchmarks, even if they fail on others. This is because, according to these authors, "positive results from model learners are more meaningful than negative results" (p. 21). Since LLMs are trained on plain text without "grounding", i.e., information available to a child learner in a physical and social learning environment, positive results from such ablated conditions provide a compelling baseline of what can be learned without domain-specific principles and constraints of human language.

But it is facile to assert the utility of grounding information as if its absence necessarily places LLMs at a disadvantage. Theoretically, enriching the text with additional semantic/pragmatic information does not necessarily enhance learnability (Niyogi 2006). More empirically, it is not clear whether grounding, to the extent it helps children, can similarly benefit LLMs. On the one hand, the environmental input arrives in many shapes and flavors: they often come into conflict, with differential effects on child learners across developmental stages. While social and communicative cues such as eye-gaze do provide guidance for language acquisition among toddlers (Baldwin 1991), they fail to draw attention away from perceptually privileged cues (e.g., bright colors, objects in motion) during infancy (e.g., Pruden et al. 2006). On the other hand, many or perhaps most elements in language (e.g., abstract words such as *of, can, good, time, think,* etc.) are not plausibly grounded in the world, yet children can learn them extremely early. And there is evidence that the richness of the sensory environment may overwhelm the learner (Gillette et al. 1999). Furthermore, language acquisition is remarkably resilient even when grounding is severely limited. The empirical literature is vast: blind children learn language in ways comparable to sighted children (Landau and Gleitman 1985), home signers spontaneously create a combinatorial syntactic system (Goldin-Meadow and Yang 2017), and perhaps most dramatically, a tactile linguistic system can be established via touch for deaf-blind individuals (Chomsky 1986). Of course, none of this suggests that children do not make use of grounding information, just as no one has suggested that language acquisition takes place in a vacuum. But children make use of such information in highly specific ways that cannot be divorced from developmental factors, linguistic and otherwise: there is no guarantee that LLMs can benefit similarly. In fact, there is evidence that they do not: augmenting LLMs with visuo-spatial cues fails to improve learning performance (Yun et al. 2021), and an ablation experiment of multimodal learning fails to yield a detectable advantage of grounding (Amariucai and Warstadt 2024).

If the potential benefit of grounding information is an over-estimation, the actual benefit of the LLM training scheme over child learners has gone largely uncommented. Like most machine learning models, LLMs operate by iterative optimization over the training data. But there is no psychological evidence for such a learning process in children. If anything, empirical research has uncovered severe limitations of children's memory, attention, and other computational capacities. Even in the simplest word learning task, children can only focus on a single referent meaning even though multiple are available in the environment (Woodard et al. 2016). Five-year-olds have difficulty integrating contextual considerations in sentence comprehension and show little or no ability to recover from parsing ambiguities (Trueswell et al. 1999). In artificial language learning experiments, children generally only learn a single linguistic form even though the input by design contains variability (Hudson Kam and Newport 2009). And a long-established experimental methodology exploits children's very cognitive limitations: failure to imitate adult sentences is taken as an indication of underdeveloped knowledge of particular syntactic structures (McNeill 1970). In fact, research from several different traditions (Elman 1993, Newport 1990, Yang 2016) suggests that restrictions on children's computational capacity are beneficial for the acquisition of language.

Taken together, LLMs and human children operate under significantly different learning conditions. In the concluding remarks of this paper, we will put forward some additional considerations from child language that ought to be connected with LLM modeling. As it stands, however, LLMs and language acquisition are like apples and oranges: even if LLMs were to achieve perfectly human-like linguistic ability, they still stand to tell us nothing about the nature of language and cognition.

Our main concern in this paper, however, is narrower and more immediate. We ask: All provisos aside, have LLMs delivered an existence proof as claimed? As usual, the proof is in the pudding. To invoke the familiar goals of linguistics (Chomsky 1965), the question of explanatory adequacy is only on the table when descriptive adequacy has been achieved.

We contend that LLMs are still far from delivering an existence proof: their relevance as cognitive models of language acquisition is therefore a moot point. Our position is informed by previous efforts, individually and collectively, to investigate the validity of models of language and cognition, especially the need for well-chosen baseline, or null hypothesis, comparisons. In particular, we have found it useful to evaluate language learning proposals against n-grams, the simplest statistical model possible. For example, the apparent success of a Bayesian learning model (Yang and Piantadosi 2022) is undercut by the fact that it is matched by a provably inadequate trigram model of language (Kodner et al. 2022). In the arena of LLMs, proposed tests for hierarchical syntactic structures (Van Schijndel and Linzen 2018, Prasad et al. 2019) are in fact not syntactic at all: not only does a definitionally linear n-gram model succeed at the tests, but so does an LLM when the words in the test sentences are randomly scrambled (Kodner and Gupta 2020). The current paper builds on the earlier work of Vázquez Martínez (2021), the first to make use of datasets that more realistically reflect human language users' syntactic knowledge for the purposes of testing LLMs' abilities: LLMs offer scant improvement over, once again, n-gram models. Overall, our past experience is concordant with findings in NLP, that LLMs often take unexpected shortcuts, exploiting unforeseen biases in the data and evaluation methods (e.g., Chao et al. 2018; McCoy et al. 2019; Wang et al. 2022). That is, LLMs may succeed at tasks in ways that are misleading and unintended by the task designers. Positive results do not necessarily demonstrate a proof-of-concept.

In this chapter,[1] we evaluate LLMs as models of language acquisition on two benchmarking data sets: the widely used Benchmark of Linguistic Minimal Pairs (BLiMP; Warstadt et al. 2020), which also forms part of the evaluation for the BabyLM Challenge, and Zorro (Huebner et al. 2021), a data set inspired by BLiMP for the express purpose of testing LLMs as proxies for child language acquisition with restricted vocabulary and trained only on child-directed speech (CDS). Section 2 reviews some key features of child language acquisition and methodological issues in the assessment of syntactic knowledge. In Section 3, we introduce the BLiMP and Zorro benchmarks and subject them to baseline tests by simple non-human-like models. These establish several weaknesses in the organization and content of both benchmarks. In Section 4, we evaluate neural models on a more challenging data set derived directly from papers in theoretical linguistics. We find that LLMs are not necessarily human-like in terms of within- and across-model variability. Finally, Section 5 concludes with a discussion of the logical problem of behavioral probing. We argue for benchmarks that better probe the structural knowledge of syntax—and more and better baseline models. We also

[1] Our evaluation code and data are available at https://github.com/hjvm/benchmarking_acquisition

make recommendations for future work on LLM modeling, especially with respect to developmental findings from child language acquisition.

2. Grammaticality, acceptability and language acquisition

One of the goals of linguistic theory is to characterize the properties that distinguish grammatical from ungrammatical sentences in a language. Grammaticality is most saliently observed in minimally contrasting examples: *colorless green ideas sleep furiously* and *furiously sleep ideas green colorless* form the most prominent pair (Chomsky 1957). In practice, the empirical study of grammaticality typically, and certainly most conveniently, relies on native speakers' acceptability judgments. Such practice is not without complications. Acceptability judgment reflects the complex interaction between the grammar and other cognitive and perceptual systems; it therefore generally produces gradient responses as opposed to the binary outcome of grammaticality. For example, longer and more complex sentences, even when fully grammatical, are rated as less acceptable than shorter and simpler sentences. Nevertheless, large-scale investigations have established the structural basis of a categorical grammar (Sprouse and Hornstein 2013). For example, syntactic constraints that prohibit certain transformational processes are shown to have a "super-additive" effect beyond acceptability degradation due to sentence length and other non-structural factors. Furthermore, acceptability judgment data collected at scale are highly consistent with the data reported in the theoretical literature, typically gathered informally with few consultants (Sprouse and Almeida 2012; Sprouse et al. 2013; Sprouse and Almeida 2017).

Unsurprisingly, by far the most widely used benchmark for assessing LLMs is also based on acceptability discrimination (Linzen et al. 2016a; Chowdhury and Zamparelli 2018; Gulordava et al. 2018; Wilcox et al. 2018; McCoy et al. 2020; Hu et al. 2020; Warstadt et al. 2020; Papadimitriou et al. 2021; Huebner et al. 2021; Sinclair et al. 2022), mirroring the practice in the empirical study of language. In this sense, LLMs are a continuation of classical statistical language models such as *n*-grams. Statistical language models are trained to assign probabilities to strings: minimal pairs can therefore be submitted for assessment in humans and machines alike when machines assign a higher probability to acceptable strings than to unacceptable ones. To this end, large acceptability rating benchmarks have been developed in various forms for syntax, semantics, and morphology: these include BLiMP (Warstadt et al. 2020), SyntaxGym (Gauthier et al. 2020), and CoLA (Warstadt et al. 2019), which sit alongside smaller-scale studies focused on specific linguistic phenomena (e.g., Linzen et al. 2016b;

Marvin and Linzen 2018; Wilcox et al. 2018; Yedetore et al. 2023). A wide range of positive results has been reported, which has led to claims of LLMs as informative cognitive models of language.

Before we examine these benchmarks and their suitability for LLM evaluation, it is once again useful to consider how children acquire stable grammars reflected in the consistency in acceptability studies. Recent years have seen (renewed) interest in individual differences across child learners (Kidd et al. 2018), especially with respect to vocabulary acquisition (Frank et al. 2021). Longitudinal records of child language development have made it possible to track both children's vocabulary growth and the development of the structural aspects of their grammar. In the Providence Corpus (Demuth et al. 2006), for example, six children were recorded at regular intervals from age one to three. On average, fewer than 20% of the first 100 words are shared between any two children. The overlap merely rises to about 40% for the first 1,000, which is the upper limit of a three-year-old's vocabulary size (Hart and Risley 1995; Bornstein et al. 2004). Yet these children's grammars are highly uniform even at this early stage. Major syntactic categories, word order and argument structure, and the core morphological rules are firmly established before age three (Brown 1973) on the basis of at most around 10 million words per year (Hart and Risley 1995) and a vocabulary size of only a few hundred types (Fenson et al. 1994). Recent research has also seen a convergence between the formal and psychological study of language and the quantitative study of language variation, use, and change. Labov, for instance, often speaks of "the enigma of uniformity:" It is remarkable that the grammars acquired by individuals in a speech community show "a high degree of uniformity in both the categorical and variable aspects of language production, where individual variation is reduced below the level of linguistic significance" (2012; see also Labov 1972).

The acquisition of vocabulary and grammar provides guidelines for the investigation of LLMs, and indeed all models of language learning. Vocabulary acquisition is a matter of rote learning. This includes not just the arbitrary pairing of sounds and meanings, but also morphological processes (e.g., irregularity) and syntactic structures (e.g., sub-categorization, collocations, etc.). There is no escape from experience: more data does result in better learning. But the structural aspects of the grammar are different. They require form generalizations over the vocabularies: as noted above, very different vocabularies apparently yield identical grammars.

The distinction between rote learning and structural learning—roughly, words vs. rules—is not well reflected by existing LLM benchmarks, including those discussed in this paper. In practice, these benchmarks are a mixture of tests for both vocabulary learning and grammar learning. Moreover, many benchmarks are stochastically generated by templates, which come with their own host of problems. While this practice makes a large number of sentences immediately available for testing—and statistically significant results may be obtained—this comes at the expense of structural diversities that are particularly revealing and the potential for compounding biases stemming from how the templates are designed. We return to this issue with concrete examples in Section 3.

Furthermore, the sentences used in LLM benchmarks are often highly unnatural and semantically and pragmatically uncontrolled. This is precisely the confounding factor that linguists seek to neutralize when attempting to uncover the structural basis of language. Authors such as Warstadt et al. (2020) are aware of the unnaturalness of their generated sentences. They present this example: *Sam ran around some glaciers*. Nevertheless, these researchers do not regard such semantic and pragmatic oddities as a problem, arguing that since they design sentence pairs to be minimally lexically distinct, they affect both sentences in a test pair equally. This reasoning relies on the (untested) assumptions that the injection of noise does not obscure the signal and that it does not affect ratings of acceptable and unacceptable sentences more than the other. However, we cannot rule out unintended, and indeed unknown, interactions between grammaticality and felicity in acceptability judgment. In fact, Sprouse et al. (2018) find that semantic implausibility affects sentence well-formedness results, even in forced choice tasks, conceptually similar to the LLM benchmark tasks. At the same time, we *have* seen case studies where deliberate mutilation of the signal—random permutation of words in a sentence (Kodner and Gupta 2020) —does not impede model performance in categorizing sentence types: the purported structural signal cannot be the thing, or at least the only thing, that models in the study captured.

Fortunately, there is already a large amount of material that not only reflects the realistic complexity of syntax but has also been carefully curated to minimize the confound of non-grammatical factors. Furthermore, these datasets, such as the LI-Adger dataset discussed in Section 4, have been evaluated for acceptability at an individual level by a large pool of native speaker subjects and show very high convergence rates across individuals. If

LLMs can truly capture the structural regularities of syntax, they should exhibit a similar degree of uniformity.

3. Re-examining the benchmarks

3.1 BLiMP (Warstadt et al. 2020)

Warstadt et al. (2020) introduce the Benchmark of Linguistic Minimal Pairs (BLiMP)[2] as a means of evaluating the linguistic knowledge of neural language models. BLiMP extends the reasoning of earlier studies (e.g., Linzen et al. 2016b; Marvin and Linzen 2018; Wilcox et al. 2018), which use a minimal pair paradigm to approximate acceptability judgments. Instead of prompting for acceptability judgments on individual sentences, as is most commonly done with human subjects, they present a model, LLM or otherwise, with two sentences that only differ in one structural or lexical property, as in (1). For a given minimal pair m_i consisting of an acceptable sentence $s_{i,1}$ (labeled in the data set as *sentence_good*) and an unacceptable sentence $s_{i,2}$ (labeled as *sentence_bad*), if a model evaluates $P(s_{i,1}) > P(s_{i,2})$, then it has succeeded on m_i. A model is scored according to the percentage of all the minimal pairs for which it assigned a higher probability to the acceptable sentence. The minimal pair approach allows for the direct evaluation of LLMs without training a binary classifier on top of them, as was necessary for previous acceptability benchmarks (e.g., CoLA; Warstadt et al. 2019).

(1) Sentence pair from BLiMP's adjunct_island phenomenon

sentence_good: Who should Derek hug after shocking Richard?

sentence_bad: Who should Derek hug Richard after shocking?

Minimal pairs need to be carefully constructed to control for length and lexical frequencies. BLiMP aims to accomplish this with automatic generation from templates, but as we discuss, this comes with the side-effect of low structural diversity and implausible semantics. The benchmark corpus includes data sets for 12 linguistic *phenomena*, including Anaphor agreement, Argument structure, Binding, Control/raising, among others listed in Tables 3-4. These are further divided into 67 *paradigms*, each containing 1,000 sentences pairs, which are meant to test variants of the phenomena. For example, the phenomenon Determiner-noun agr. contains six paradigms for adjacent agreement, agreement with irregular nouns, and agreement with adjectives

---

[2] https://github.com/alexwarstadt/blimp

intervening. BLiMP has become a standard NLP benchmark for this task and was used as part of the test data for the 2023 BabyLM Challenge.

3.2 Zorro (Huebner et al. 2021)

Huebner et al. (2021) aim to explicitly evaluate the relationship between LLMs and the acquisition of grammar. They introduce BabyBERTa, "an acquisition-friendly version of RoBERTa," which Portelance and Jasbi (2023) describe as a "proof-of-concept" model. Variants of BabyBERTa are trained using only CDS from ao-childes (Huebner and Willits 2021; BabyBERTa_ ao-childes), a pre-processed version of English CHILDES (MacWhinney 1991), as well as variants trained on larger datasets from other sources.

Because BabyBERTa_ao-childes (henceforth BabyBERTa) was trained on much less text than RoBERTa and related models are, its vocabulary is much smaller. To mitigate the impact of out-of-vocabulary items on their tests, the authors introduce a new grammaticality test suite, Zorro,[3] in the style of BLiMP (2). Sentence pairs are generated for one paradigm each for 11 of BLiMP's 12 phenomena, along with two additional phenomena. While Zorro bills itself as a benchmark as similar to BLiMP as possible given the constraints on its vocabulary, we show that the Zorro sentences are not only lexically simpler as intended, but their templates are also far less complex and even less varied than the sentences in the corresponding BLiMP phenomena. Full lists of paradigms for each data set can be found in Table 7.2, and the full data sets themselves are made available by the benchmarks' original authors.

(2) Sentence pair from Zorro's local_attractor-in_question_with_aux phenomenon

sentence_good: is the whale getting the person?

sentence_bad: is the whale gets the person?

3.3 Linear baselines

As noted earlier, BLiMP and Zorro tests are automatically generated from category-based templates. This way, a large number of examples can be collected and tested, but the drawback is that all examples are essentially the same structure. Moreover, many of the structures are simple, falling considerably below the coverage of modern syntactic analyses. In fact, many phenomena appear solvable by strictly linear methods *in practice*. Some of

[3] https://github.com/phueb/Zorro/

these do not even require structural knowledge *in principle*, and instead test lexical memorization. The observation that such template-generated examples can be solved this way is not new to the field. For example, Kam et al. (2008) demonstrated that a bigram model will predict the grammatical sentence from template-produced pairs featuring auxiliary inversion (a structural phenomenon) about as well as neural models of the time.

Consider BLiMP's ANAPHOR AGREEMENT paradigm. This tests whether a model recognizes agreement between an anaphor (e.g., *himself / herself*) and its antecedent, but there are two problems here. First, the antecedent is always the first and only noun, preceding the anaphor in the sentence (3). A linear rule like "find the leftmost noun, then make sure it matches the anaphor" should be able to solve it. This is a weakness in the template used to generate these minimal pairs.

(3) Sentence pair from BLiMPs's anaphor_number_agreement phenomenon

sentence_good: Susan revealed herself.

sentence_bad: Susan revealed themselves.

Second, the task relies heavily on lexical knowledge. Since the antecedent in most of the minimal pairs is an English proper name, the test reduces to memorizing which gender is conventionally assigned to the names (4). Other sentences incorporate common nouns and *itself* and *themselves*, but these too reduce to identifying human animate vs. non-human and singular vs. plural nouns. Furthermore, some minimal pairs, as in (5), are invalid. Obviously, both *himself* and *herself* could agree with *dancer* in the absence of any disambiguating context, so both sentences in the pair are grammatical.

(4) Sentence pair from BLiMPs's anaphor_gender_agreement phenomenon

sentence_good: Katherine can't help herself.

sentence_bad: Katherine can't help himself.

(5) Invalid sentence pair from BLiMPs's anaphor_gender_agreement phenomenon *sentence_good*: That dancer wouldn't aggravate herself.

sentence_bad: That dancer wouldn't aggravate himself.

The Subject-verb agr phenomenon provides another example of systemic problems. This phenomenon is widely used in behavioral probing because noun phrases intervening between the subject and verb can serve as agreement distractors. These are useful in distinguishing whether a model is attempting to enforce agreement with the closest noun or with the structural subject. Consider (6) from Linzen et al. (2016b, (9)), which uses such sentences as probes for LLM structural knowledge:

(6) The **roses** in the *vase* by the *door* **are** red.

The bold verb should agree with the bold subject plural noun, and not the intervening italicized singular nouns. While such an agreement pattern does not necessarily require a structural rule (all of Linzen et al.'s examples require agreement with the first/leftmost noun, a linear pattern), it does require a model to rely on a long-distance dependency. However, BLiMP's phenomenon only partially evaluates this. It is made up of six paradigms, four of which only evaluate string-adjacent agreement with no distractor, for example (7). In the two paradigms with intervening distractor nouns, the target for agreement is always the first noun, for example (8), so a linear rule along the lines of "always agree with the first/leftmost noun" would completely suffice to solve the entire phenomenon.

(7) Sentence pair from BLiMPs's regular_plural_subject_verb_agreement _1 phenomenon

sentence_good: Most legislatures haven't disliked children.

sentence_bad: Most legislatures hasn't disliked children.

(8) Sentence pair from BLiMPs's distractor_agreement_relational_noun phenomenon *sentence_good*: A niece of most senators hasn't descended most slopes. *sentence_bad*: A niece of most senators haven't descended most slopes.

Anaphor and Subject-verb agr, among others, as implemented, are just lexical and string-based phenomena – they do not actually test a model's syntactic knowledge. In principle, an *n*-gram model with sufficiently large *n* and enough training data to capture the conventional gender of English names should be able to solve them perfectly. Success on such simple tests tells us little about the genuine grammatical capacity of LLMs and distorts or dilutes summary metrics calculated over the benchmark. We evaluate this concern quantitatively with two studies of linear rules that do not incorporate structural knowledge. We find that many paradigms are indeed solvable with non-human-like linear approaches. These paradigms, therefore, do not contribute to the overall goal of evaluating whether an LLM possesses linguistic knowledge. Additionally, we find that the paradigms of Zorro tend to be structurally even simpler and less internally varied than the parallel paradigms of BLiMP. It is a weaker benchmark even when accounting for the intended lexical simplicity.

3.3.1 N-Gram models

The original BLiMP paper reports the accuracy of a 5-gram model trained on the 3.1 billion token Gigaword Corpus (Graff et al. 2003), in addition to the accuracy of three neural LLMs and humans. They find that the 5-gram model scores above chance (>50%) on all but two phenomena and even surpasses at least one neural model on 25 of the 67 paradigms and four of the 12 phenomena. We point out that one could therefore conclude, by the logic of the test, that the 5-gram model has "learned" a substantial amount about the linguistic phenomena, even though we know that the 5-gram model is just a linear string model with a fixed context window. Surprisingly, this was not the authors' conclusion. Despite the 5-gram model's substantially above-chance performance, they conclude "The 5-gram model's poor performance – overall and on every individual category – indicates that BLiMP is likely not solvable from local co-occurrence statistics alone" (Warstadt et al. 2020, §5).

A more detailed look at the Gigaword 5-gram model's performance proves insightful. In general, and as expected, it struggles with paradigms with long-distance dependencies that stretch the limits of its small context window. Revisiting Subject-verb agr. for illustration, its performance is only slightly weaker than the LLMs on each of the four string-adjacent paradigms, while it performs much worse, and well below chance, on the two paradigms with distractors. This makes sense. Since the distractor nouns lay within the 5-gram model's context window and the true subject often did not, the model tended to prefer agreement between the distractor and verb. It should be noted that the neural models also perform substantially worse (though still well above chance) on the distractor paradigms than on the string-adjacent paradigms.

However, the 5-gram model also performs well on some paradigms that it should not be able to. It outperforms two or three neural models on four of eight Island paradigms, for example. Islands are a syntactic phenomenon *par excellence*, so this result suggests a problem with the data. At the very least, phenomena on which the 5-gram model readily scores above chance should be removed from evaluation, since they are not informative about a model's structural knowledge.

Since our focus is primarily on language acquisition, and the Gigaword 5-gram is trained on far too much data to serve as a reasonable model in that setting, we conduct our own tests with a 5-gram model with back-off trained only on ao-childes. We compare these results to BabyBERTa, the acquisition-relevant derivative of RoBERTa trained on the same data.[4] A 5-gram model trained on only ao-childes is expected to be excessively limited in its vocabulary. It is likely to be missing test *n*-grams present in both benchmarks, which is a particular problem, since this type of model has no way to generalize over lexical categories. To give the linear baseline an additional chance to succeed while adding minimal complexity, we evaluate both a 5-gram word model (5-word) and a 5-gram part-of-speech (POS) model trained only on POS tags (5-tag). To train the POS model, ao-childes was tagged using GPoSTTL, a rule-based POS tagger with a tokenizer and lemmatizer based on the Brill Tagger (Brill 1992). This was used to train the sklearn package's CRF POS-tagger, which was then used to label the benchmark corpora. This approach of training the 5-tag model only on labeled ao-childes was taken to avoid bringing in additional knowledge from a tagger trained on larger corpora. The downside is that the tagger is not accurate in labeling the ungrammatical benchmark sentences, which likely hurts the performance of the 5-tag model. Nevertheless, we observe that the 5-tag model often performs well even in spite of this inherent limitation. Our use of POS is motivated by a developmental perspective. Syntactic categories can be formed purely distributionally as early as infancy (Mintz 2003; Shi and Melançon 2010; Reeder et al. 2013), and children almost never make mistakes in their use of syntactic categories (Valian 1986). It is thus plausible to assume that the acquisition of grammatical knowledge builds on a developmentally prior stage of syntactic category learning.

In addition to the 5-word and 5-tag models, we evaluate an oracle that marks a correct prediction if either 5-word or 5-tag makes a correct prediction on a given minimal pair. The oracle is not in itself a fair direct comparison, because it requires access to the minimal pair labels to determine whether or not either 5-gram model answered correctly. Rather, it should be interpreted as a summary metric for the relationship between 5-word and 5-tag. An oracle score only slightly above that of the two 5-gram models would indicate that they tend to succeed and fail on mostly the same minimal pairs, while a high relative oracle score indicates that the two 5-gram models tend to make complementary errors instead. That is, errors are not necessarily attributable

---

[4] We downloaded the publicly available model checkpoints from the BabyBERTa GitHub repository and replicated the BLiMP and Zorro results hosted on the Zorro GitHub repository.

to the string-local limitations of 5-grams *per se* but rather to 5-gram sparsity or errors in tagging. This is what we observe.

The results of the 5-gram experiments are summarized in Table 7.1 with breakdowns by Zorro and BLiMP paradigm in Tables 7.2-7.4. We draw three conclusions from these. First, the 5-gram models perform surprisingly well relative to the BabyBERTa transformer despite their extremely non-human-like simplicity when trained on the same ao-childes data. Either 5-word or 5-tag outperformed BabyBERTa on 11 of 23 Zorro paradigms and 21 of 67 BLiMP paradigms. As expected, the ao-childes 5-gram models perform worse on BLiMP than the Gigaword 5-gram model, but so does BabyBERTa, and both still achieve high accuracy on several paradigms scattered across the phenomena.

Second, the 5-gram oracle outperforms 5-word, 5-tag, and BabyBERTa on a majority of paradigms and on overall accuracy on both BLiMP and Zorro. This means that the 5-word and 5-tag models tended to succeed on different minimal pairs. The high oracle score is another sign that the paradigms often capture surface properties rather than structural properties that would stump 5-gram models.

Third, the 5-gram models outperform BabyBERTa on proportionately more Zorro paradigms than BLiMP paradigms. Additionally, the ao-childes 5-word model achieved 78.91% performance on Zorro, while the Gigaword 5-gram model only reached 60.5% on BLiMP. If Zorro preserved BLiMP's complexity and merely accounted for the smaller vocabulary in the ao-childes training data, we should expect much more similar performance on both of these measures. Taken together, these suggest that Zorro is a substantially weaker benchmark than BLiMP, and it more greatly overestimates the apparent positive results of the acquisition-inspired BabyBERTa.

Table 7.1: Summary performance for 5-grams relative to BabyBERTa on Zorro and BLiMP. The number of paradigms in which a 5-gram model outperforms BabyBERTa and the overall average accuracy across paradigms are reported. Either = either 5-word or 5-tag outpeformed BabyBERTa on the entire paradigm. Oracle = sentence pairs were marked correct if either 5-word or 5-tag made the correctprediction.

Zorro	BabyBERTa	5-Word	5-Tag	Either	Oracle
# Best	–	8/23	8/23	11/23	14/23
Avg Acc	78.91%	63.44%	57.59%	–	83.43%

BLiMP	BabyBERTa	5-Word	5-Tag	Either	Oracle
# Best	–	18/67	10/67	23/67	48/67
Avg Acc	60.72%	50.72%	37.93%	–	68.32%

Table 7.2: Word and tag-level 5-gram models trained on ao-childes plus 5-Gram Oracle and Simple Linear Rule Oracle for Zorro. 5-Gram and Simple Rule scores that are greater than BabyBERTa_ ao-childes scores are bolded.

Phenomenon	Paradigm	BabyBERTa AO-CHILDES	5-Gram			Simple Rule
			Word	Tag	Oracle	
agreement_ subject_verb	across_rel_clause	64.85	50.95	46.35	68.95	96.20
	in_simple_question	92.35	61.15	90.9	93.9	98.30
	in_question_with_aux	90.85	59	80.15	90.9	98.05
	across_prep_phrase	72.85	50	50	62.6	98.40
agreement_ determiner_ noun	between_neighbors	91.3	83.05	49.85	88.6	98.60
	across_1_adjective	89.85	50.45	50.05	75.05	97.20
filler-gap	wh_question_object	98.75	42.8	100	100	100
	wh_question_subject	75.7	88.3	76.55	97.1	100
island-effects	coord_struct_constr	97.05	43.35	55.6	83.85	100
	adjunct_island	56.15	66.1	58.8	83.85	100
quantifiers	existential_there	92.9	80.25	38.4	89.55	100
	superlative	64.55	45.1	82	96.05	100
npi_licensing	only_npi_licensor	74.1	79.4	3.7	79.4	100
	matrix_question	65.25	47.5	28.65	58	100
argument_ structure	swapped_arguments	91	92.15	81.7	98.85	100
	transitive	60.05	64.15	32.65	78.6	58.05
	dropped_argument	79.9	85.05	83.6	95.75	100
irregular	verb	69.65	62.9	93.6	96.35	88.40
anaphor_ agreement	pronoun_gender	51.75	49.15	1.95	50.95	52.75
ellipsis	n_bar	55.3	66.6	63.6	89.9	100
binding	principle_a	89.4	45.9	3.6	47.75	100
case	subjective_pronoun	94.7	99.55	97.95	100	100
local_attractor	in_question_with_aux	96.65	55.65	95	99.05	100
AVERAGE		78.91%	63.44 %	57.59 %	83.43%	93.97%
Fraction ≥ BabyBERTa		–	8/23	8/23	14/23	22/23

Table 7.3: Word and tag-level 5-gram models trained on AO-CHILDES plus 5-Gram Oracle and Simple Linear Rule Oracle for BLiMP (ANAPHOR AGR through ELLIPSIS). 5-Gram and Simple Rule scores that are greater than BabyBERTa_AO-CHILDES scores are bolded.

Phenomenon	Paradigm	BabyBERTa AO-CHILDES	5-Gram			Simple Rule
			Word	Tag	Oracle	
ANAPHOR AGR	anaphor_gender_agreement	65.6	26.3	8	33.9	73.9
	anaphor_number_agreement	73.7	52.9	5.7	55.5	80.1
ARG STRUCTURE	animate_subject_passive	45.5	48.4	24	58.4	65.25
	animate_subject_trans	59.7	50	57.1	78.2	84.65
	causative	58.5	55.2	30.7	68.8	85.6
	drop_argument	63.2	50.9	52.9	80.8	77.1
	inchoative	50.7	56	37.1	73.8	57.1
	intransitive	52.1	48.2	49.6	76.3	73.55
	passive_1	50.2	52.1	12.9	56.4	59.5
	passive_2	54	48.4	18.1	56.8	59.6
	transitive	55.3	51.6	36.1	67.6	57.85
BINDING	principle_A_case_1	43.6	100	7.1	100	100
	principle_A_case_2	99.9	41.5	13	48.3	99.2
	principle_A_c_command	58.7	35.7	4.2	38.1	71.35
	principle_A_domain_1	96.5	38.4	3.1	40.7	100
	principle_A_domain_2	51.4	61.7	2.7	62.8	58.3
	principle_A_domain_3	46.8	44.5	29.7	61.1	50.4
	principle_A_reconstruction	40.9	32.1	53.9	68	74.1

CONTROL/ RAISING	existential_there_ obj.	59.1	30.5	23.4	46.5	67.95
	existential_there_ subj.	51	43.4	17	53.6	77
	expletive_it_ object_raising	63.3	61.2	48.3	79.6	69.5
	tough_vs_raising_ 1	72.2	59.1	49.6	83.2	87.1
	tough_vs_raising_ 2	34.4	41.3	18.4	54.1	92.5
DET-NOUN AGR	agr_irregular_1	66.6	48.8	37.4	61.3	68.45
(agreement = agr)	agr_irregular_2	87.4	74.3	12.3	77.1	73.7
	agr_with_ adjective_1	76.3	48.2	49.7	63.8	95.95
	agr_with_adj_ irregular_1	82.9	49	49.7	56.3	74.45
	agr_with_adj_ irregular_2	67	49.5	18.3	58.2	71.8
	agr_with_adj_2	80.4	49.8	19.9	59.7	95.6
	agr_1	72.2	64.1	48.1	74.5	95.55
	agr_2	87.4	65.2	11	68.1	96.75
ELLIPSIS	ellipsis_n_bar_1	58.7	64.1	63.5	86.4	85.65
	ellipsis_n_bar_2	42.8	39.9	70.5	80.9	99.95

Table 7.4: Word and tag-level 5-gram models trained on AO-CHILDES plus 5-Gram Oracle and Simple Linear Rule Oracle for BLiMP (FILLER-GAP through SUBJECT-VERB AGR). 5-Gram and Simple Rule scores that are greater than BabyBERTa_AO-CHILDES scores are bolded.

Phenomenon	Paradigm	BabyBERTa	5-Gram			Simple
		AO-CHILDES	Word	Tag	Oracle	Rule
FILLER-GAP	wh_questions_object_ gap	73	37	82.4	89.2	99.95
	wh_questions_subject _gap	79.9	49	81.4	89.4	99.9

	wh_vs_that_no_gap	90.9	77.2	83.8	94.9	99.95
	wh_vs_that_no_gap_long_dist.	92.1	74.9	87	95.8	99.7
	wh_vs_that_with_gap	29.1	22.7	15	33	100
	wh_vs_that_with_gap_long_dist.	14.9	25.8	12.8	32.8	99.9
IRREGULAR FORMS	irregular_past_participle_adj.	59.8	99.4	12.2	99.4	100
	irregular_past_participle_verbs	59.8	99.4	12.2	99.4	100
ISLAND EFFECTS	adjunct_island	63.8	58.4	55.5	82.5	94.5
(coordinate_structure_	csc_complex_left_branch	36.2	11.8	19.6	26.9	97.05
constraint = csc)	csc_object_extraction	56.5	41.9	37.1	63.7	86.35
	left_branch_island_echo_question	52.4	16.3	30.1	38.7	100
	left_branch_island_simple_question	66.6	24.5	30.3	43.8	97.9
	sentential_subject_island	46.1	37.3	42.8	62.9	82.65
	wh_island	47.1	69	93.4	97.3	100
NPI LICENSING	matrix_question_npi_licensor_pres.	56.4	41.1	39.5	65.7	97.4
(sent = sentential)	npi_present_1	27	56	26.7	69.6	100
	npi_present_2	20.3	56.4	25.8	70.5	100
	only_npi_licensor_pres.	71.6	98.4	2.4	98.5	100
	only_npi_scope	72.1	80.4	79.4	97.2	100
	sent_negation_npi_licensor_pres.	73.8	100	0	100	100
	sent_negation_npi_scope	81.9	40	65.3	79.6	100

QUANTIFIERS	existential_there_ quantifiers_1	93.7	79.1	26.4	87.4	97.3
	existential_there_ quantifiers_2	35.7	19.6	36	50.6	96.85
	superlative_ quantifiers_1	49.5	73	89.8	96.4	100
	superlative_ quantifiers_2	61.2	51.9	0.1	52	100
SUBJECT-VERB AGR	distractor_agr_ relational_noun	29	26.2	21.4	42.1	50.25
(agr = agreement)	distractor_agr_relative _clause	35.6	28.3	30.4	49.8	55.85
	irregular_plural_ subject_verb_agr_1	67.9	33.4	51.7	62.5	53.2
	irregular_plural_ subject_verb_agr_2	66.2	51	51.9	70.7	59.3
	regular_plural_ subject_verb_agr_1	68.8	39.9	51.1	72	64.35
	regular_plural_ subject_verb_agr_2	60.1	51	55.6	76.9	73.15
AVERAGE across Table 3 and 4		60.72%	50.72%	37.93%	68.32%	84.35%
Fraction ≥ BabyBERTa		–	18/67	10/67	48/67	62/67

3.3.2 Hand-written simple rules

In addition to reporting results on 5-gram models, we created simple hand-written rules, inspired by the pervasive "agree with the first/leftmost noun" pattern, which demonstrate that the probes are solvable in principle without reference to linguistic structure. While we do not claim that such rules are akin to the state of knowledge in LLMs, it is also difficult to completely rule out this possibility. On the one hand, it is still unclear how to interpret the representation of linguistic knowledge in LLMs. On the other hand, the vast majority of training data, at least child-directed for language acquisition, is structurally simple and can in fact be handled by rule-like pattern matchers. In English CDS, the distribution of anaphora is exceedingly straightforward: almost all instances of *himself* are preceded in the sentence by the subject pronoun *he* and a (male) noun phrase with no co-referential competitors. For comparison, Zorro adjunct_island can be solved perfectly by always selecting the sentence where the third-last word is *the*, and many of the paradigms can be solved by tracking the index of a specific word. Others can be solved by checking for the presence of a certain word. For example, the superlative paradigm can be solved by accepting the sentence that contains either *more* or *fewer*. If any of these patterns hold in the training data, then there is a real possibility that the models have learned and are exploiting them. For both Zorro and BLiMP, the same rule can often be used to solve more than one paradigm. We write simple linear rules for each Zorro and BLiMP paradigm, summarized in Tables 7.5-7.7.

Table 7.5: Simple Linear Rule descriptions for Zorro. Rules that require sentences within a minimal pair to be compared are marked with an asterisk (in_question_with_aux).

Phenomenon	**Paradigm**	**Rule Description**
agreement_	across_rel_clause	2nd word ends in *s* iff 3rd last is in {*are,*
subject_verb	in_simple_question	*were, do*}
		Word 2 right of {*are, were*} ends in *s*. Word
agreement_	in_question_with_aux	2 right of {*is, are*} does not
determiner_noun	across_prep_phrase	4th word ends in *s* iff 2nd is in {*are, were,*
	between_neighbors	*do*} 2nd word ends in *s* iff 3rd last is in
FILLER-GAP		{*are, were, do*} If {*these, those*} in
island-effects	across_1_adjective	sentence, next word ends in *s*. If {*this,*
		that} in sentence, next word does not
QUANTIFIERS	wh_question_object	If {*these, those*} in sentence, word 2 right
npi_licensing	wh_question_subject	ends in *s*. If {*this, that*} in sentence, word
argument_structure	coord_struct_constr	2 right does not 2nd word is *the*
	adjunct_island	*who* does not immediately precede *the*
	existential_there	4th word is *and*
irregular	superlative	3rd last word is *the*
anaphor_agreement	only_npi_licensor	Contains one of {*many, some, no, few, a,*
ellipsis	matrix_question	*an*} Contains one of {*more, fewer*}
binding	swapped_arguments	1st word is *only*
case	transitive	Contains one of {*does, will, should, could,*
	dropped_argument	*did, wouldo*} 1st word is *the*
	verb	2nd last word does not end in *e*
	pronoun_gender	1st word is *the*
	n_bar	word following *had* ends in *n* or no word
	principle_a	ends in *n*
	subjective_pronoun	Sentence contains *himself*
	local_attractor	*Sentence where *and* appears farther
	in_question_with_aux	rightward 4th last word ends with *ing*
		1st word is *the*
		4th word does not end with *'s*

Table 7.6: Linear Rule descriptions for BLiMP (ANAPHOR AGR through ELLIPSIS). Rules that require sentences to be compared are marked with an asterisk.

Phenomenon	Paradigm	Rule Description
ANAPHOR	anaphor_gender_agreenment	Does not contain *itself*
	anaphor_number_agreenment	Number of words that end in *s* is even
ARG STRUCTURE	animate_subject_passive	Contains one of {*men, woman, children, teacher, lad, offspring, student,customer, girl, boy*}
	animate_subject_trans	*Is the shorter of the two sentences
	causative transitive	Does not contain one of {*appear, vanish, exist, sigh, rush, cheer, clash, fall, fell, waste*}
	drop_argument inchoative intransitive	Last word is not one of {*to, with, about, from, at, through, by, like*}
	passive_1 passive_2	None of {*communicat, suffer, compet, shout, laugh, scream, complain, compromis, grin, chat*} in sentence
BINDING	principle_A_case_1	*Is the shorter of the two sentences
	principle_A_case_2	*Is the longer of the two sentences
	principle_A_c_command	(Last word ends in *s* and (sl word is any of pl_det or the 2nd word is *lot*)) or 2nd to last word ends in *s*
	principle_A_domain_1	*Is the shorter of the two sentences
	principle_A_domain_2	*Is the shorter of the two sentences
	principle_A_domain_3	Does not contain *that*
	principle_A_reconstruction	4th word does not end in *ed* nor *'t*
CONTROL/ RAISING	existential_there_obj	Does not contain one of {*ask, press, entic, prod, obligat, convinc, badger, compel, sway, order*}
	existential_there_subj.	Contains one of {*certain, soon, likely, unlikely, bound, about*} or {*appear, sure, threaten, look*}

	expletive_it_object_raising	Does not contain one of {*ask, press, entic, prod, obligat, convinc, badger, compel, sway, order*}
	tough_vs_raising_1	Does not contain one of {*certain, soon, likely, unlikely, bound, about*}, nor *apt*
	tough_vs_raising_2	Contains one of {*certain, soon, likely, unlikely, bound, about*} or *apt*
DET-NOUN AGR (agreement = agr)	agr_irregular_1 agr_irregular_2 agr_with_adj_irregular_1 agr_with_adj_irregular_2	Does not end in *that* followed by (one of {*people, women, men, children*} or a word ending in *ses*) nor in {*those, these*} followed by (a word ending in *is* or not with *s* at all)
	agr_with_adjective_1 agr_with_adj_2 agr_1 agr_2	Does not end in *that* followed by a word ending in a letter other than *i* followed by *s* nor in {*those, these*} followed by (a word ending in *is* or not with *s* at all)
ELLIPSIS	ellipsis_n_bar_1	Last word in {*one-ten, many, few, several, more, some, lot, fewer*}
	ellipsis_n_bar_2	Last word has already occurred in sentence

Table 7.7: Linear Rule descriptions for BLiMP (Filler-gap through Subject-verb agr). Rules that require sentences to be compared are marked with an asterisk.

Phenomenon	Paradigm	Rule Description
FILLER-GAP	wh_questions_object_gap	Does not contain *wh*
	wh_questions_subject_gap	Does not contain *wh*
	wh_vs_ that_no_gap	Does not contain *wh*
	wh_vs_that_no_gap_long_dist.	Does not contain *wh*
	wh_vs_that_with_gap	Contains *wh*
	wh_vs_that_with_gap_long_dist.	Contains *wh*
IRREGULAR FORMS	irregular_past_participle_adj.	If 1st word is *the*, then 2nd word ends in *n*, otherwise 2nd word must not end in *n*
	irregular _past_participle_verbs	*Is the shorter of the two sentences

ISLAND EFFECTS (coordinate_structure_constraint = csc)	adjunct_island	Last word is not *about* and does not end in *ing*
	csc_complex_left_branch	2nd word is not in {*had, should, is, was, can, has, will, would, could, do, does, might, were, did*}
	csc_object_extraction	2nd to last word is not *and*
	left_branch_island_echo_question	Does not start with *Wh*
	left_branch_island_simple_question	2nd word is not in {*had, should, is, was, can, has, will, would, could, do, does, might, were, did*}
	sentential_subject_island	Ends in *ing* or *ed* or *with*
	wh_island	*wh*, capitalized or not, occurs twice in sentence
NPI LINCENSING (sentential=sent)	matrix_question_npi_licensor _pres.	1st word in {*had, should, is, was, can, has, will, would, could, do, does, might, were, did*}
	npi_present_1	Does not contain the word *ever*
	npi_present_2	Does not contain the word *ever*
	only_npi_licensor_pres.	1st word is *only*
	only_npi_scope	1st word is *only*
	sent_negation_npi_licensor _pres.	Does not contain the word *ever*
	sent_negation__npi_scope	Does not contain the word *ever*
QUANTIFIERS	existential_there_quantifiers_ 1 existential_there_quantifiers_2	Does not contain {*each, most, all, every*} while also containing {*one-ten*}
	superlative_quantifiers_ 1	*Is the longer of the two sentences
	superlative_quantifiers_2	1st word is not *no*
SUBJECT-VERB AGR (agreement = agr)	distractor _agr_relational_noun	*Is the longer of the two sentences
	distractor_agr_relative_clause irregular_plural_subject_ verb_agr _1	The number of words that ends in *s* is odd

	irregular_plural_ subject_verb_agr _2	Contains no word ending in a letter other than *i* and followed by *s* that is followed by a word ending in *s*
	regular_plural_ subject_verb_agr_1	None of {*people, women, men, children*} is followed by a word ending in *s*
	regular_plural_ subject_verb_agr _2	*Is the shorter of the two sentences The number of words that ends in *s* is odd

In summary, the hand-written rules yielded 93.97% accuracy on Zorro and solved 14 of 23 Zorro paradigms with 100% accuracy. Each agreement_ paradigm is solved with at least 96% accuracy, where the 4% loss is completely attributable to the presence of two irregular nouns, *feet* and *children,* in some of the minimal pairs, which do not end in the -*s* referenced by these rules. The paradigm principle_A_case_2 is solved only with 99.2% accuracy because there are several test pairs where the grammatical sentence is actually a copy of the ungrammatical sentence (e.g., *Leslie imagined herself skated around the hospital*), while all valid grammatical sentences contain a word ending in -*ing*. The lowest performance is 52.75% on anaphor_agreement-pronoun_gender, a paradigm that requires a model to "know" the canonical gender of English names in order to choose *himself* or *herself* (the same problem is exhibited in BLiMP's equivalent (4)). The test sentence pairs were not quite balanced, so always guessing *himself* instead of *herself* yields just over 50% accuracy.

BLiMP proved more challenging. Our rules only yielded 84.35% accuracy on average and achieved perfect scores on 14 of 67 rules. Yet, this is still a high score. The general success of the handwritten simple linear rules suggests that BLiMP suffers from the same issues regarding lack of sentence variety that Zorro does, but the lower accuracy indicates that the problem is not as severe. In principle, we could have composed more complex rules that achieved perfect accuracy on all paradigms; however, these simpler rules better illustrate our points. The success of non-human-like simple linear rules on most paradigms of both benchmarks further emphasizes that success on the template-based behavioral task does not necessarily imply that an LLM possesses linguistic knowledge.

4. An alternative benchmark: The LI-Adger dataset

The LI-Adger dataset is a comprehensive collection of 519 sentence types, 300 collected by Sprouse et al. (2013) from *Linguistic Inquiry (LI) 2001-2010*,[5] a

[5] https://www.jonsprouse.com/data/Lingua2013/SSA.data.zip

major theoretical journal in linguistics, and 219 collected by Sprouse and Almeida (2012) from Adger 2003 *Core Syntax* textbook.[6] Each sentence type includes eight hand-constructed sentences, assembled into 150 pairwise (LI) and 105 multi-condition (Adger) phenomena where each minimal pair is lexically matched. We provide an example from each dataset in Table 7.8.

While the LI-Adger dataset serves as a behavioral probe, and is thus subject to the inherent weaknesses of such approaches, it improves upon prior template-based benchmark data sets in three key ways. First, unlike BLiMP and Zorro, the LI-Adger sentences were hand-constructed by theoretical linguists who are well aware of the contribution of non-grammatical factors such as lexical frequency, sentence parsing strategies, and semantic and pragmatic plausibility to the behavioral measure of acceptability rating (Sprouse et al. 2018). Therefore, best efforts have been made to control for these confounds and zero in on the structural properties of the grammar. The strong confirmation from large-scale judgment data provides methodological support for such practices, as well as the empirical status of these datasets. Second, the 255 total pairwise and multi-condition phenomena achieve much more diverse coverage of syntactic phenomena than the 67 paradigms in BLiMP and the 23 paradigms in Zorro. They were not generated from templates nor even devised by a single research team, since they were collected from across the literature. Third, the human judgments were collected using the Magnitude Estimation (ME) task (and Likert Scale (LS) in the case of the LI sentences) in addition to the Forced-Choice (FC) paradigm that was used for the BLiMP human baselines. This is a crucial advantage because the FC task treats sentence acceptability as functionally categorical: A sentence is only acceptable or not relative to its minimal pair counterpart, whereas tasks such as ME allow us to make comparisons within and across minimal pairs, thereby treating sentence acceptability as a gradient measure.

With this dataset, we conduct the following two tests. First, in line with Vázquez Martínez (2021), we sort the LI-Adger dataset into 2,391 unique minimal pairs and collect pseudo log-likelihood scores for each sentence from several models evaluated by Huebner et al. (2021). We score them using the same criteria as BLiMP and Zorro, where a model is expected to assign a higher probability to the grammatical member of a grammatical-ungrammatical sentence pair. As a baseline for the models, we include Log-Likelihood and Syntactic Log-Odds Ratio (SLOR; Pauls and Klein 2012; Lau et al. 2017) scores

[6] https://www.jonsprouse.com/data/JoL2012/SA2012.data.zip

by a trigram model trained on the British National Corpus (BNC; 100M words) by Sprouse et al. (2018).

We include the results of this test in Figure 7.1. We observe that all models are further from the human baseline as compared to those in BLiMP (no human baselines were reported for Zorro). But more importantly, we observe that the trigram model scored using SLOR performs on par with the BabyBERTa models and approaches the performance of RoBERTa (Liu et al. 2019) trained on 10 million words. If we were to adopt Warstadt and Bowman's (2022) "positive results from model learners are more meaningful than negative results" argument, then the trigram model is as suitable a model of language acquisition as BabyBERTa is.

Figure 7.1: LM performance on the LI-Adger dataset. Human performance is marked by the vertical line. Baby=BabyBERTa, CHI=AO-CHILDES, News=AO-NEWSELA, Wiki=Wikipedia-1.

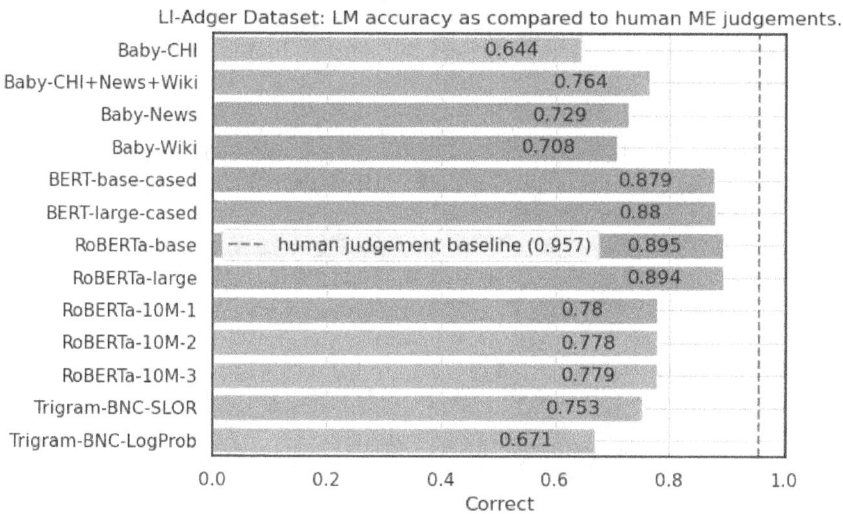

Raw accuracy notwithstanding, we proceed to conduct a novel test of judgment variability on our collection of LLMs. We take advantage of the structure of the LI-Adger dataset in the following way: There are 519 sentence types, and for each type, there are eight sentences that retain the same syntactic structure but vary lexical items at the locus of the syntactic structure tested (e.g., the head of a verb phrase from which extraction takes place). These datasets thus allow us to contrast the consistency of human judgment across and within construction types against that of the LMs.

We z-score the LLM and trigram model judgments to make them comparable to the human judgments. Then, for each set of eight sentences, we take the mean and standard deviation of all the judgments for humans and each model. We find that the models are much more variable in their judgments: The human judgments, on average, vary by 0.288 standard deviation (std. dev.) units within a given set of sentences. On the other hand, the LLM that varies is BabyBERTa Wiki, varying by 0.451 std. dev. units on average. The rest of the models nearly double the variability of the human judgments, ranging from 0.518 for RoBERTa-10M-1 to 0,554 for BERT-large-cased. Variability appears to increase rather than decrease as training size and performance increase. Surprisingly, the trigram model, when scored using log probabilities, is the closest in variability to the human judgments at 0.331 std. dev. units, but also the furthest when scored using SLOR with a variability of 0.599. Once again, we find that a positive result on one test or another is not enough to draw positive conclusions. This within-model variability provides another data point in addition to raw accuracy, which can be compared between people and machines, one that casts LLMs in a less human-like light.

Figure 7.2: Correlation matrices of human judgments and LM output means (top) and standard deviations (bottom) on each sentence type on the LI-Adger dataset. Baby=BabyBERTa, CHI=AO-CHILDES, News=AO-NEWSELA, Wiki=Wikipedia-1.

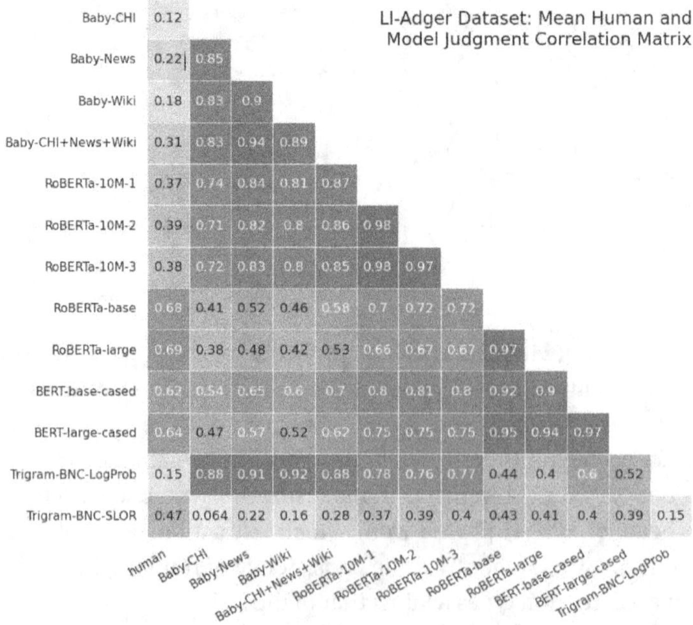

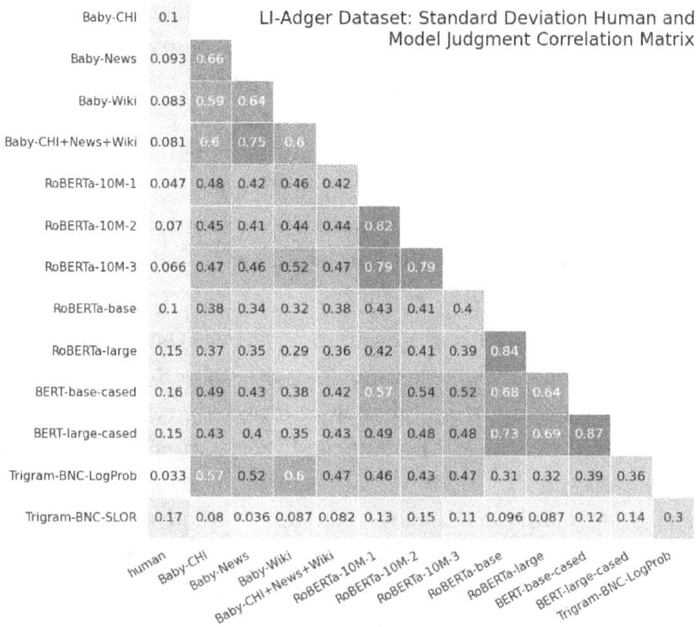

Exploring this further, we correlate the means and standard deviations of 512 sentence types across each LM and humans and plot the results in Figure 7.2. Both in terms of mean and standard deviation, we observe generally high correlations between the various LLMs, but much lower correlations between the LLMs and humans. As humans are the odd ones out, this suggests that whatever the LLMs learn does not appear to be human-like. Interestingly, the BabyBERTa LLMs show very high correlations with the naive trigram log-likelihood scores and very low correlations with trigram SLOR scores, raising further suspicions that these small acquisition-inspired LLMs behave like a very non-human-like model.

5. Conclusions and prospects

The development of LLMs has generated remarkable technological advances and intellectual excitement. But a clear-eyed assessment of their capacity and limitations is still necessary, not least because of the closed nature of most of the leading LLM models, and the fact that there is very little theoretical understanding of how and why these models work.

The claim of LLMs as an existence proof for language acquisition appears premature. The scientific relevance of LLMs is predicated upon, among others, the assumption that despite the differences in the learning mechanisms and learning environment, LLMs already demonstrate human-like linguistic abilities. By critically revisiting the commonly used evaluation benchmarks, we contend that current LLMs still have some ways to go. The inaccurate assessment of LLMs stems from the weakness of existing benchmarks, which can be met in unintended ways and, in any case, do not reflect a representative range of linguistic phenomena. Furthermore, we suggest that binarized tasks be complemented with professionally curated datasets that include gradient acceptability judgments from human subjects.

Our study is about the limitations of evaluation, so it is useful to point out the limitations of our study as well. Most obviously, ours and any study would benefit from testing and reporting on a wider range of neural models and a wider range of baselines: we have only considered very simple *n*-gram language models, while the classic NLP toolbox contains many others. For example, the problem of aligning words and images is a well-studied machine learning problem: think Google image search. It would be informative to compare classic non-neural methods against LLMs in the modeling of lexical acquisition (Vong et al. 2024). And like most work in this area, our evaluations were only performed in English. Fortunately, recent years have seen the development of LI-Adger-type datasets for specific phenomena (Sprouse and Hornstein 2013) as well as for other languages (Kush et al. 2019; Chen et al. 2020; Fukuda et al. 2022; Al-Aqarbeh and Sprouse 2023). All behavioral probes, of course, should be submitted for rigorous assessment.

We end with some broader discussion about language acquisition and the cognitive interpretation of computational models such as LLMs. While it is now widely recognized that children learn language with only a fraction of the data needed for LLM training, merely reducing the amount of training data alone—such as the 100-million-word threshold in the BabyLM Challenge—still falls far short of the requirement for an adequate model of language acquisition. It is true that a native speaker's knowledge of language can be established on the basis of approximately 100 million words, but child language research makes clear that not all aspects of linguistic knowledge are learned at the same time: the developmental trajectory of language acquisition matters as much as the final grammar attained (Pinker 1984; Yang 2002). Some, such as inflectional morphology, case marking, word order, and major transformations are acquired very early in all languages studied so far (e.g., Brown 1973; Slobin 2022) at an order of magnitude fewer words of input, while others are learned rather late: These include derivational morphology (Jarmulowicz 2002), control and cleft structures (Chomsky 1969) and the dative constructions (Gropen et al. 1989) in the case of English, but these may emerge much earlier in other languages. For example, English-learning children use the passive structure relatively late (Borer and Wexler 1987; Pinker et al. 1987) but no such delay is found in the passives of Sesotho (Demuth 1989) and Inuktitut (Allen and Crago 1996) learning children.

Successful learning in the "limit" (e.g., 100 million words), therefore, is not an inadequate measure of LLMs as cognitive models of language acquisition. In this sense, the current discussion of LLMs and their cognitive relevance compares unfavorably against the earlier connectionist approach to language, which was deeply concerned with the empirical study of child language and subsequently led to many empirical discoveries (McClelland and Patterson 2002, Pinker and Ullman 2002). Let us not forget that the major thrust of the original PDP model for morphological learning (Rumelhart and McClelland 1986) was an account of the abrupt emergence of productivity and over-regularization (Ervin and Miller 1963, Brown 1973), previously attributed to symbolic rules. Disappointingly, child language development has received virtually no attention in the LLM literature. A few case studies, however, are already informative. While a neural model of English past tense (Kirov and Cotterell 2018) eventually learns the "add -ed" rule, it does so with over 3,000 verb lemmas. By contrast, children learn that rule before or around 3 (Kuczaj 1977), when their vocabulary only contains around 300 or so verbs. At the same time, child language acquisition often exhibits non-monotonic learning trajectories, with important implications for learning theories: once again, the phenomenon of over-regularization in English past tense, but similar patterns are found in phonological, syntactic, and semantic acquisition (e.g., Richter 2021; Bowerman 1988; Crain and Thornton 2000). However, to the best of our knowledge, LLM learning shows no similar developmental trends as performance monotonically improves as the training data size increases (Belth et al. 2021; Zhang et al. 2021; Kodner et al. 2023).

In our understanding, the research goals of the LLM community in fact align well with the developments in theoretical linguistics in the past thirty years, as both aim to minimize principles and constraints unique to language (Chomsky 1995, Hauser et al. 2002, Berwick and Chomsky 2016). In our own work, again individually and collectively, we have developed distributional learning models as alternative accounts to a domain-specific Universal Grammar (Yang 2002, 2016; Li et al. 2021; Belth et al. 2021; Kodner 2022; Heuser et al. 2024). These efforts may find useful points of contact with LLMs, so long as the goals of descriptive and explanatory adequacy are broadly shared and respected.

Bibliography

Adger, David. 2003. *Core Syntax: A Minimalist Approach.* Oxford: Oxford University Press. https://doi.org/10.1093/oso/9780199243709.001.0001

Al-Aqarbeh, Rania, and Jon Sprouse. 2023. "Island Effects and Amelioration by Resumption in Jordanian Arabic: An Auditory Acceptability-Judgment Study." *Syntax.* https://doi.org/10.1111/synt.12262

Allen, Shanley E. M., and Martha B. Crago. 1996. "Early Passive Acquisition in Inuktitut." *Journal of Child Language* 23 (1): 129–55. https://doi.org/10.1017/S0305000900010126

Amariucai, Theodor, and Alex Warstadt. 2024. "Acquiring Linguistic Knowledge from Multimodal Input." In *Proceedings of the 27th Conference on Computational Natural Language Learning: The BabyLM Challenge,* 128–41. https://doi.org/10.18653/v1/2023.conll-babylm.11

Baldwin, Dare A. 1991. "Infants' Contribution to the Achievement of Joint Reference." *Child Development* 62: 875–90. https://doi.org/10.1111/j.1467-8624.1991.tb01577.x

Belth, Caleb, Sarah Payne, Deniz Beser, Jordan Kodner, and Charles Yang. 2021. "The Greedy and Recursive Search for Morphological Productivity." In *Proceedings of CogSci 2021.*

Berwick, Robert, and Noam Chomsky. 2016. *Why Only Us: Language and Evolution.* Cambridge, MA: MIT Press. https://doi.org/10.7551/mitpress/9780262034241.001.0001

Borer, Hagit, and Kenneth Wexler. 1987. "The Maturation of Syntax." In *Parameter Setting,* edited by Thomas Roeper and Edwin Williams, 123–72. Dordrecht: Reidel. https://doi.org/10.1007/978-94-009-3727-7_6

Bornstein, Marc H., Linda R. Cote, Sharone Maital, et al. 2004. "Cross-Linguistic Analysis of Vocabulary in Young Children: Spanish, Dutch, French, Hebrew, Italian, Korean, and American English." *Child Development* 75 (4): 1115–39. https://doi.org/10.1111/j.1467-8624.2004.00729.x

Bowerman, Melissa. 1988. "The 'No Negative Evidence' Problem: How Do Children Avoid Constructing an Overly General Grammar?" In *Explaining Language Universals,* edited by John A. Hawkins, 73–101. Oxford: Basil Blackwell.

Brill, Eric. 1992. "A Simple Rule-Based Part of Speech Tagger." In *Proceedings of the Third Conference on Applied Natural Language Processing,* 152–55. Stroudsburg, PA: Association for Computational Linguistics. https://doi.org/10.3115/974499.974526

Brown, Roger. 1973. *A First Language: The Early Stages.* Cambridge, MA: Harvard University Press. https://doi.org/10.4159/harvard.9780674732469

Chao, Wei-Lun, Hexiang Hu, and Fei Sha. 2018. "Being Negative but Constructively: Lessons Learnt from Creating Better Visual Question

Answering Datasets." In *Proceedings of the 2018 Conference of the North American Chapter of the Association for Computational Linguistics: Human Language Technologies, Volume 1 (Long Papers)*, 431–41. Stroudsburg, PA: Association for Computational Linguistics. https://doi.org/10.18653/v1/N18-1040

Chen, Zhong, Yuhang Xu, and Zhiguo Xie. 2020. "Assessing Introspective Linguistic Judgments Quantitatively: The Case of the Syntax of Chinese." *Journal of East Asian Linguistics* 29: 311–36. https://doi.org/10.1007/s10831-020-09210-y

Chomsky, Carol. 1969. *The Acquisition of Syntax in Children from 5 to 10.* Cambridge, MA: MIT Press.

Chomsky, Carol. 1986. "Analytic Study of the Tadoma Method: Language Abilities of Three Deaf-Blind Subjects." *Journal of Speech, Language, and Hearing Research* 29 (3): 332–47. https://doi.org/10.1044/jshr.2903.347

Chomsky, Noam. 1957. *Syntactic Structures.* The Hague: Mouton. https://doi.org/10.1515/9783112316009

Chomsky, Noam. 1965. *Aspects of the Theory of Syntax.* Cambridge, MA: MIT Press. https://doi.org/10.21236/AD0616323

Chomsky, Noam. 1995. *The Minimalist Program.* Cambridge, MA: MIT Press.

Chowdhury, Shammur Absar, and Roberto Zamparelli. 2018. "RNN Simulations of Grammaticality Judgments on Long-Distance Dependencies." In *Proceedings of the 27th International Conference on Computational Linguistics*, 133–44. Stroudsburg, PA: Association for Computational Linguistics.

Crain, Stephen, and Rosalind Thornton. 2000. *Investigations in Universal Grammar: A Guide to Experiments on the Acquisition of Syntax and Semantics.* Cambridge, MA: MIT Press. https://doi.org/10.7551/mitpress/3948.001.0001

Demuth, Katherine. 1989. "Maturation and the Acquisition of the Sesotho Passive." *Language* 65 (1): 56–80. https://doi.org/10.2307/414842

Demuth, Katherine, Jennifer Culbertson, and Jennifer Alter. 2006. "Word-Minimality, Epenthesis, and Coda Licensing in the Acquisition of English." *Language and Speech* 49 (2): 137–73. https://doi.org/10.1177/00238309060490020201

Elman, Jeffrey L. 1993. "Learning and Development in Neural Networks: The Importance of Starting Small." *Cognition* 48 (1): 71–99. https://doi.org/10.1016/0010-0277(93)90058-4

Ervin, Susan M., and Wick R. Miller. 1963. "Language Development." In *Child Psychology: The Sixty-Second Yearbook of the National Society for the Study of Education*, edited by Harold Stevenson, 108–43. Chicago: University of Chicago Press. https://doi.org/10.1037/13101-004

Fenson, Larry, Philip S. Dale, J. Steven Reznick, Elizabeth Bates, Donna J. Thal, Stephen J. Pethick, Michael Tomasello, Carolyn B. Mervis, and Joan Stiles. 1994. "Variability in Early Communicative Development." *Monographs of the Society for Research in Child Development* 59 (5): 1–185. https://doi.org/10.2307/1166093

Frank, Michael C., Mika Braginsky, Daniel Yurovsky, and Virginia A. Marchman. 2021. *Variability and Consistency in Early Language Learning: The Wordbank Project.* Cambridge, MA: MIT Press. https://doi.org/10.7551/mitpress/11577.001.0001

Fukuda, Shin, Nozomi Tanaka, Hajime Ono, and Jon Sprouse. 2022. "An Experimental Reassessment of Complex NP Islands with NP-Scrambling in Japanese." *Glossa: A Journal of General Linguistics* 7 (1). https://doi.org/10.16995/glossa.5737

Gauthier, Jon, Jennifer Hu, Ethan Wilcox, Peng Qian, and Roger Levy. 2020. "SyntaxGym: An Online Platform for Targeted Evaluation of Language Models." In *Proceedings of the 58th Annual Meeting of the Association for Computational Linguistics: System Demonstrations,* 70–76. Stroudsburg, PA: Association for Computational Linguistics. https://doi.org/10.18653/v1/2020.acl-demos.10

Gillette, Jane, Henry Gleitman, Lila Gleitman, and Anne Lederer. 1999. "Human Simulations of Vocabulary Learning." *Cognition* 73 (2): 135–76. https://doi.org/10.1016/S0010-0277(99)00036-0

Goldin-Meadow, Susan, and Charles Yang. 2017. "Statistical Evidence That a Child Can Create a Combinatorial Linguistic System without External Linguistic Input: Implications for Language Evolution." *Neuroscience and Biobehavioral Reviews* 81 (Part B): 150–57. https://doi.org/10.1016/j.neubiorev.2016.12.016

Graff, David, Junbo Kong, Ke Chen, and Kazuaki Maeda. 2003. "English Gigaword." *Linguistic Data Consortium, Philadelphia* 4 (1): 34.

Gropen, Jess, Steven Pinker, Michelle Hollander, Richard Goldberg, and Ronald Wilson. 1989. "The Learnability and Acquisition of the Dative Alternation in English." *Language* 65 (2): 203–57. https://doi.org/10.2307/415332

Gulordava, Kristina, Piotr Bojanowski, Edouard Grave, Tal Linzen, and Marco Baroni. 2018. "Colorless Green Recurrent Networks Dream Hierarchically." In *Proceedings of the 2018 Conference of the North American Chapter of the Association for Computational Linguistics: Human Language Technologies, Volume 1 (Long Papers),* 1195–1205. Stroudsburg, PA: Association for Computational Linguistics. https://doi.org/10.18653/v1/N18-1108

Hart, Betty, and Todd R. Risley. 1995. *Meaningful Differences in the Everyday Experience of Young American Children.* Baltimore: Paul H. Brookes Publishing.

Hauser, Marc D., Noam Chomsky, and W. Tecumseh Fitch. 2002. "The Faculty of Language: What Is It, Who Has It, and How Did It Evolve?" *Science* 298 (5598): 1569–79. https://doi.org/10.1126/science.298.5598.1569

Heuser, Annika, Héctor Vázquez Martínez, and Charles Yang. 2024. "The Learnability of Syntactic Islands." In *Proceedings of the 54th Northeast Linguistic Society Meeting.* Forthcoming.

Hu, Jennifer, Jon Gauthier, Peng Qian, Ethan Wilcox, and Roger Levy. 2020. "A Systematic Assessment of Syntactic Generalization in Neural Language

Models." In *Proceedings of the 58th Annual Meeting of the Association for Computational Linguistics*, 1725–44. Stroudsburg, PA: Association for Computational Linguistics. https://doi.org/10.18653/v1/2020.acl-main.158

Hudson Kam, Carla L., and Elissa L. Newport. 2009. "Getting It Right by Getting It Wrong: When Learners Change Languages." *Cognitive Psychology* 59 (1): 30–66. https://doi.org/10.1016/j.cogpsych.2009.01.001

Huebner, Philip A., Elior Sulem, Cynthia Fisher, and Dan Roth. 2021. "BabyBERTa: Learning More Grammar with Small-Scale Child-Directed Language." In *Proceedings of the 25th Conference on Computational Natural Language Learning*, 624–46. Stroudsburg, PA: Association for Computational Linguistics. https://doi.org/10.18653/v1/2021.conll-1.49

Huebner, Philip A., and Jon A. Willits. 2021. "Using Lexical Context to Discover the Noun Category: Younger Children Have It Easier." In *Psychology of Learning and Motivation* 75: 279–331. https://doi.org/10.1016/bs.plm.2021.08.002

Jarmulowicz, Linda D. 2002. "English Derivational Suffix Frequency and Children's Stress Judgments." *Brain and Language* 81 (1–3): 192–204. https://doi.org/10.1006/brln.2001.2517

Kam, Xuân-Nga Cao, Iglika Stoyneshka, Lidiya Tornyova, Janet D. Fodor, and William G. Sakas. 2008. "Bigrams and the Richness of the Stimulus." *Cognitive Science* 32 (4): 771–87. https://doi.org/10.1080/03640210802067053

Kidd, Evan, Seamus Donnelly, and Morten H. Christiansen. 2018. "Individual Differences in Language Acquisition and Processing." *Trends in Cognitive Sciences* 22 (2): 154–69. https://doi.org/10.1016/j.tics.2017.11.006

Kirov, Christo, and Ryan Cotterell. 2018. "Recurrent Neural Networks in Linguistic Theory: Revisiting Pinker and Prince (1988) and the Past Tense Debate." *Transactions of the Association for Computational Linguistics* 6: 651–65. https://doi.org/10.1162/tacl_a_00247

Kodner, Jordan. 2022. "What Learning Latin Verbal Morphology Tells Us about Morphological Theory." *Natural Language and Linguistic Theory* 41: 733–92. https://doi.org/10.1007/s11049-022-09553-2

Kodner, Jordan, Spencer Caplan, and Charles Yang. 2022. "Another Model Not for the Learning of Language." *Proceedings of the National Academy of Sciences* 119 (29): e2204664119. https://doi.org/10.1073/pnas.2204664119

Kodner, Jordan, and Nitish Gupta. 2020. "Overestimation of Syntactic Representation in Neural Language Models." In *Proceedings of the Annual Meeting of the Association for Computational Linguistics (ACL)*, 1757–62. Stroudsburg, PA: Association for Computational Linguistics. https://doi.org/10.18653/v1/2020.acl-main.160

Kodner, Jordan, Salam Khalifa, Sarah Payne, and Zoey Liu. 2023. "Re-Evaluating the Evaluation of Neural Morphological Inflection Models." In *Proceedings of the 45th Annual Meeting of the Cognitive Science Society (CogSci)* 45: 3259–67.

Kuczaj, Stan A. 1977. "The Acquisition of Regular and Irregular Past Tense Forms." *Journal of Verbal Learning and Verbal Behavior* 16 (5): 589–600. https://doi.org/10.1016/S0022-5371(77)80021-2

Kush, Dave, Terje Lohndal, and Jon Sprouse. 2019. "On the Island Sensitivity of Topicalization in Norwegian: An Experimental Investigation." *Language* 95 (3): 393–420. https://doi.org/10.1353/lan.2019.0051

Labov, William. 1972. *Sociolinguistic Patterns.* Philadelphia: University of Pennsylvania Press.

Labov, William. 2012. "What Is to Be Learned." *Review of Cognitive Linguistics* 10 (2): 265–93. https://doi.org/10.1075/rcl.10.2.02lab

Landau, Barbara, and Lila R. Gleitman. 1985. *Language and Experience: Evidence from the Blind Child.* Cambridge, MA: Harvard University Press.

Lau, Jey Han, Alexander Clark, and Shalom Lappin. 2017. "Grammaticality, Acceptability, and Probability: A Probabilistic View of Linguistic Knowledge." *Cognitive Science* 41 (5): 1202–24. https://doi.org/10.1111/cogs.12414

Li, Daoxin, Lydia Grohe, Petra Schultz, and Charles Yang. 2021. "The Distributional Learning of Recursive Structures." In *Proceedings of the 45th Boston University Conference on Language Development,* 471–85. Somerville, MA: Cascadilla Press. https://doi.org/10.31234/osf.io/xhz4m

Linzen, Tal, and Marco Baroni. 2021. "Syntactic Structure from Deep Learning." *Annual Review of Linguistics* 7: 195–212. https://doi.org/10.1146/annurev-linguistics-032020-051035

Linzen, Tal, Emmanuel Dupoux, and Yoav Goldberg. 2016a. "Assessing the Ability of LSTMs to Learn Syntax-Sensitive Dependencies." *Transactions of the Association for Computational Linguistics* 4: 521–35. https://doi.org/10.1162/tacl_a_00115

Linzen, Tal, Emmanuel Dupoux, and Yoav Goldberg. 2016b. "Assessing the Ability of LSTMs to Learn Syntax-Sensitive Dependencies." *Transactions of the Association for Computational Linguistics* 4: 521–35. https://doi.org/10.1162/tacl_a_00115

Liu, Yinhan, Myle Ott, Naman Goyal, Jingfei Du, Mandar Joshi, Danqi Chen, Omer Levy, Mike Lewis, Luke Zettlemoyer, and Veselin Stoyanov. 2019. "RoBERTa: A Robustly Optimized BERT Pretraining Approach." *arXiv* preprint arXiv:1907.11692.

MacWhinney, Brian. 1991. *The CHILDES Language Project: Tools for Analyzing Talk.* Hillsdale, NJ: Lawrence Erlbaum.

Marvin, Rebecca, and Tal Linzen. 2018. "Targeted Syntactic Evaluation of Language Models." In *Proceedings of the 2018 Conference on Empirical Methods in Natural Language Processing,* 1192–202. Stroudsburg, PA: Association for Computational Linguistics. https://doi.org/10.18653/v1/D18-1151

McClelland, James L., and Karalyn Patterson. 2002. "Rules or Connections in Past-Tense Inflections: What Does the Evidence Rule Out?" *Trends in Cognitive Sciences* 6 (11): 465–72. https://doi.org/10.1016/S1364-6613(02)01993-9

McCoy, R. Thomas, Robert Frank, and Tal Linzen. 2020. "Does Syntax Need to Grow on Trees? Sources of Hierarchical Inductive Bias in Sequence-to-Sequence Networks." *Transactions of the Association for Computational Linguistics* 8: 125–40. https://doi.org/10.1162/tacl_a_00304

McCoy, Tom, Ellie Pavlick, and Tal Linzen. 2019. "Right for the Wrong Reasons: Diagnosing Syntactic Heuristics in Natural Language Inference." In *Proceedings of the 57th Annual Meeting of the Association for Computational Linguistics*, 3428–38. Stroudsburg, PA: Association for Computational Linguistics. https://doi.org/10.18653/v1/P19-1334

McNeill, David. 1970. *The Acquisition of Language: The Study of Developmental Psycholinguistics.* New York: Harper and Row.

Mintz, Toben H. 2003. "Frequent Frames as a Cue for Grammatical Categories in Child Directed Speech." *Cognition* 90 (1): 91–117. https://doi.org/10.1016/S0010-0277(03)00140-9

Newport, Elissa L. 1990. "Maturational Constraints on Language Learning." *Cognitive Science* 14 (1): 11–28. https://doi.org/10.1207/s15516709cog1401_2

Niyogi, Partha. 2006. *The Computational Nature of Language Learning and Evolution.* Cambridge, MA: MIT Press. https://doi.org/10.7551/mitpress/2024.001.0001

Papadimitriou, Isabel, Ethan A. Chi, Richard Futrell, and Kyle Mahowald. 2021. "Deep Subjecthood: Higher-Order Grammatical Features in Multilingual BERT." In *Proceedings of the 16th Conference of the European Chapter of the Association for Computational Linguistics: Main Volume*, 2522–32. Stroudsburg, PA: Association for Computational Linguistics. https://doi.org/10.18653/v1/2021.eacl-main.215

Pauls, Adam, and Dan Klein. 2012. "Large-Scale Syntactic Language Modeling with Treelets." In *Proceedings of the 50th Annual Meeting of the Association for Computational Linguistics (Volume 1: Long Papers)*, 959–68. Stroudsburg, PA: Association for Computational Linguistics.

Pinker, Steven. 1984. *Language Learnability and Language Development.* Cambridge, MA: Harvard University Press.

Pinker, Steven, David S. Lebeaux, and Loren Ann Frost. 1987. "Productivity and Constraints in the Acquisition of the Passive." *Cognition* 26 (3): 195–267. https://doi.org/10.1016/S0010-0277(87)80001-X

Pinker, Steven, and Michael T. Ullman. 2002. "The Past and Future of the Past Tense." *Trends in Cognitive Sciences* 6 (11): 456–63. https://doi.org/10.1016/S1364-6613(02)01990-3

Portelance, Eva, and Masoud Jasbi. 2023. "The Roles of Neural Networks in Language Acquisition." PsyArXiv. https://doi.org/10.31234/osf.io/b697.

Prasad, Grusha, Marten van Schijndel, and Tal Linzen. 2019. "Using Priming to Uncover the Organization of Syntactic Representations in Neural Language Models." In *Proceedings of the 23rd Conference on Computational Natural Language Learning (CoNLL)*, 66–76. Stroudsburg, PA: Association for Computational Linguistics. https://doi.org/10.18653/v1/K19-1007

Pruden, Shannon M., Kathy Hirsh-Pasek, Roberta Michnick Golinkoff, and Elizabeth A. Hennon. 2006. "The Birth of Words: Ten-Month-Olds Learn Words through Perceptual Salience." *Child Development* 77 (2): 266–80. https://doi.org/10.1111/j.1467-8624.2006.00869.x

Reeder, Patricia A., Elissa L. Newport, and Richard N. Aslin. 2013. "From Shared Contexts to Syntactic Categories: The Role of Distributional Information in Learning Linguistic Form-Classes." *Cognitive Psychology* 66 (1): 30–54. https://doi.org/10.1016/j.cogpsych.2012.09.001

Richter, Caitlin. 2021. *Alternation-Sensitive Phoneme Learning: Implications for Children's Development and Language Change.* PhD diss., University of Pennsylvania.

Rumelhart, David E., and James L. McClelland. 1986. "On Learning the Past Tenses of English Verbs." In *Parallel Distributed Processing: Explorations into the Microstructure of Cognition. Volume 2: Psychological and Biological Models,* edited by James L. McClelland, David E. Rumelhart, and the PDP Research Group, 16–271. Cambridge, MA: MIT Press.

Shi, Rushen, and Andréane Melançon. 2010. "Syntactic Categorization in French-Learning Infants." *Infancy* 15 (5): 517–33. https://doi.org/10.1111/j.1532-7078.2009.00022.x

Sinclair, Arabella, Jaap Jumelet, Willem Zuidema, and Raquel Fernández. 2022. "Structural Persistence in Language Models: Priming as a Window into Abstract Language Representations." *Transactions of the Association for Computational Linguistics* 10: 1031–50. https://doi.org/10.1162/tacl_a_00504

Slobin, Dan Isaac. 2022. *The Crosslinguistic Study of Language Acquisition: Volume 3.* New York: Psychology Press. https://doi.org/10.4324/9781315808208

Sprouse, Jon, and Diogo Almeida. 2012. "Assessing the Reliability of Textbook Data in Syntax: Adger's Core Syntax." *Journal of Linguistics* 48 (3): 609–52. https://doi.org/10.1017/S0022226712000011

Sprouse, Jon, and Diogo Almeida. 2017. "Design Sensitivity and Statistical Power in Acceptability Judgment Experiments." *Glossa* 2 (1): 1. https://doi.org/10.5334/gjgl.236

Sprouse, Jon, and Norbert Hornstein. 2013. *Experimental Syntax and Island Effects.* Cambridge: Cambridge University Press.

Sprouse, Jon, Carson T. Schütze, and Diogo Almeida. 2013. "A Comparison of Informal and Formal Acceptability Judgments Using a Random Sample from *Linguistic Inquiry* 2001–2010." *Lingua* 134: 219–48. https://doi.org/10.1016/j.lingua.2013.07.002

Sprouse, Jon, Beracah Yankama, Sagar Indurkhya, Sandiway Fong, and Robert C. Berwick. 2018. "Colorless Green Ideas Do Sleep Furiously: Gradient Acceptability and the Nature of the Grammar." *The Linguistic Review* 35 (3): 575–99. https://doi.org/10.1515/tlr-2018-0005

Trueswell, John C., Irina Sekerina, Nicole M. Hill, and Marian L. Logrip. 1999. "The Kindergarten-Path Effect: Studying On-Line Sentence Processing in

Young Children." *Cognition* 73 (2): 89–134. https://doi.org/10.1016/S0010-0277(99)00032-3

Valian, Virginia. 1986. "Syntactic Categories in the Speech of Young Children." *Developmental Psychology* 22 (4): 562. https://doi.org/10.1037/0012-1649.22.4.562

Van Schijndel, Marten, and Tal Linzen. 2018. "A Neural Model of Adaptation in Reading." In *Proceedings of the 2018 Conference on Empirical Methods in Natural Language Processing*, 4704–10. Association for Computational Linguistics. https://doi.org/10.18653/v1/D18-1499

Vázquez Martínez, Héctor. 2021. "The Acceptability Delta Criterion: Testing Knowledge of Language Using the Gradience of Sentence Acceptability." In *Proceedings of the Fourth BlackboxNLP Workshop on Analyzing and Interpreting Neural Networks for NLP*, 479–95. Association for Computational Linguistics. https://doi.org/10.18653/v1/2021.blackboxnlp-1.38

Vázquez Martínez, Héctor, Annika Lea Heuser, Charles Yang, and Jordan Kodner. 2023. "Evaluating Neural Language Models as Cognitive Models of Language Acquisition." In *Proceedings of the 1st GenBench Workshop on (Benchmarking) Generalisation in NLP*, 48–64. Association for Computational Linguistics. https://doi.org/10.18653/v1/2023.genbench-1.4

Vong, Wai Keen, Wentao Wang, A. Emin Orhan, and Brenden M. Lake. 2024. "Grounded Language Acquisition through the Eyes and Ears of a Single Child." *Science* 383 (6682): 504–11. https://doi.org/10.1126/science.adi1374

Wang, Tianlu, Rohit Sridhar, Diyi Yang, and Xuezhi Wang. 2022. "Identifying and Mitigating Spurious Correlations for Improving Robustness in NLP Models." In *Findings of the Association for Computational Linguistics: NAACL 2022*, 1719–29. Association for Computational Linguistics. https://doi.org/10.18653/v1/2022.findings-naacl.130

Warstadt, Alex, and Samuel R. Bowman. 2022. "What Artificial Neural Networks Can Tell Us about Human Language Acquisition." In *Algebraic Structures in Natural Language*, 17–60. CRC Press. https://doi.org/10.1201/9781003205388-2

Warstadt, Alex, Aaron Mueller, Leshem Choshen, et al. 2023. "Findings of the BabyLM Challenge: Sample-Efficient Pretraining on Developmentally Plausible Corpora." In *Proceedings of the BabyLM Challenge at the 27th Conference on Computational Natural Language Learning*, 1–34. Association for Computational Linguistics. https://doi.org/10.18653/v1/2023.conll-babylm.1

Warstadt, Alex, Alicia Parrish, Haokun Liu, et al. 2020. "BLiMP: The Benchmark of Linguistic Minimal Pairs for English." *Transactions of the Association for Computational Linguistics* 8: 377–92. https://doi.org/10.1162/tacl_a_00321

Warstadt, Alex, Amanpreet Singh, and Samuel R. Bowman. 2019. "Neural Network Acceptability Judgments." *Transactions of the Association for Computational Linguistics* 7: 625–41. https://doi.org/10.1162/tacl_a_00290

Wilcox, Ethan, Roger Levy, Takashi Morita, and Richard Futrell. 2018. "What Do RNN Language Models Learn about Filler–Gap Dependencies?" In *Proceedings of the 2018 EMNLP Workshop BlackboxNLP: Analyzing and Interpreting Neural Networks for NLP*, 211–21. Association for Computational Linguistics. https://doi.org/10.18653/v1/W18-5423

Woodard, Kristina, Lila R. Gleitman, and John C. Trueswell. 2016. "Two- and Three-Year-Olds Track a Single Meaning during Word Learning: Evidence for Propose-but-Verify." *Language Learning and Development* 12 (3): 252–61. https://doi.org/10.1080/15475441.2016.1140581

Yang, Charles. 2002. *Knowledge and Learning in Natural Language.* Oxford: Oxford University Press.

Yang, Charles. 2016. *The Price of Linguistic Productivity: How Children Learn to Break Rules of Language.* Cambridge, MA: MIT Press. https://doi.org/10.7551/mitpress/9780262035323.001.0001

Yang, Yuan, and Steven T. Piantadosi. 2022. "One Model for the Learning of Language." *Proceedings of the National Academy of Sciences* 119 (5): e2021865119. https://doi.org/10.1073/pnas.2021865119

Yedetore, Aditya, Tal Linzen, Robert Frank, and R. Thomas McCoy. 2023. "How Poor Is the Stimulus? Evaluating Hierarchical Generalization in Neural Networks Trained on Child-Directed Speech." In *Proceedings of the 61st Annual Meeting of the Association for Computational Linguistics (Volume 1: Long Papers),* 9370–93. Association for Computational Linguistics. https://doi.org/10.18653/v1/2023.acl-long.521

Yun, Tian, Chen Sun, and Ellie Pavlick. 2021. "Does Vision-and-Language Pretraining Improve Lexical Grounding?" In *Proceedings of EMNLP.* ArXiv:2109.10246. https://doi.org/10.18653/v1/2021.findings-emnlp.370

Zhang, Yian, Alex Warstadt, Haau-Sing Li, and Samuel R. Bowman. 2021. "When Do You Need Billions of Words of Pretraining Data?" In *Proceedings of the 59th Annual Meeting of the Association for Computational Linguistics and the 11th International Joint Conference on Natural Language Processing (Volume 1: Long Papers),* 1112–25. https://doi.org/10.18653/v1/2021.acl-long.90

Chapter 8

The creative aspect of human language use: How modern language models fare with an old idea

Vincent J. Carchidi

Independent Researcher

> It is not a novel insight that human speech is distinguished by these qualities, though it is an insight that must be recaptured time and time again. With each advance in our understanding of the mechanisms of language, thought, and behavior, comes a tendency to believe that we have found the key to understanding man's apparently unique qualities of mind.
>
> Noam Chomsky, *Language and Mind.*

Abstract: Ordinary human language use is *stimulus-free, unbounded,* yet *appropriate and coherent* to the circumstances of its use, a characterization typically associated with Noam Chomsky and other generative linguists. This tripartite depiction of linguistic behavior is distinctive in that it is unique within the animal world and central to any scientific account of human language. The successes of Large Language Models (LLMs) in Natural Language Processing—chiefly in their ability to produce human-like text—altogether fail to shed any light on this "creative aspect of language use" (CALU). The chapter shows that such models fail to reproduce this ability without exception. Moreover, they offer no account of how humans acquire the ability to use their language in this tripartite fashion and what role this should play in a theory of human language. Against this background, the author argues that the failure to shed any scientific light on CALU amounts to a deficiency of a methodology that considers LLMs accurate models of human language.

Keywords: Philosophy of Science, Linguistic Theory, Large Language Models, Artificial Intelligence, Language Use, Generative Grammar

***

1. Introduction

In March 2023, Steven Piantadosi (2023) published a wide-ranging critique of Noam Chomsky's approach to linguistics.[1] "After decades of privilege and prominence in linguistics," he wrote, "Noam Chomsky's approach to the science of language is experiencing a remarkable downfall" (Piantadosi 2021, 1). The cause: computational advances embodied in contemporary Large Language Models (LLMs). LLMs, he argued, are "bona fide linguistic *theories*" (Piantadosi 2023, 7). Their ability to produce human-like grammatical utterances is considered an unprecedented simulation of human language's "high-level properties" (Piantadosi 2023, 27), made possible without invoking methodological constraints like "competence vs. performance," respect "minimality" or "perfection," and without adherence to the dictum that statistical patterns of analyzed data are to be avoided (Piantadosi, 2023, 15).

Optimism for the role of LLMs in linguistic theory—potentially as models of language in their own right—has since reached a fever pitch. This volume comprises some responses to arguments concerning their role and artificial intelligence (AI) generally in linguistics.

This chapter will likewise wade into the debate, though from a different angle. Specifically, it focuses on what Chomsky has traditionally called the "creative aspect of language use," or CALU, an old idea traceable back to Cartesian thought (Descartes 2009/1637; Cordemoy 1668). Often conflated with the recursive property of generative grammars (the ability to produce and understand an unbounded range of structured expressions), CALU denotes the full breadth of linguistic performance in Chomsky's work; the *ordinary use* of language, not the abstract knowledge it relies upon.

CALU is typically defined by Chomsky and philosophical commentary on his work as the ability to express thoughts through language over an unbounded range effectively *at will*, with no apparent reliance on local stimuli to engage in such expression, while nonetheless converging with the mental states of others who hear (or otherwise interpret) the remarks. Language use is, in this sense, *appropriate* to

[1] The preprint has since been updated and published as a chapter, referenced below.

situations but not *caused* by situations. It is operationalized according to three criteria: *stimulus-freedom*, *unboundedness*, and *appropriateness to situations* (see Chomsky 2009, ch. 1; McGilvray 2001, 6-13; Asoulin 2013, 228-232).

While there are philosophical arguments concerning the prospects for computational devices (including non-transformer based AI systems) to reproduce CALU (e.g., Carchidi 2025, Forthcoming), this chapter will instead focus on three points: First, the theoretical constraints entailed by CALU's inclusion in a science of language, namely, the invocation of a *competence-performance* distinction and an *internalist* orientation in the study of language and mind. Second, the cognitive architecture necessary to enable—but not explain—creative language use, namely, a *modular* and *generative language faculty*. Finally, it will make the case in tandem with these points that LLMs, conceived as models or theories of human language, fail to impose the requisite theoretical constraints, fail to offer a viable alternative to account for CALU other than generative linguistics, and risk excessive ambition in explanatory theory.

The chapter is structured as follows: it begins with a brief overview of the reception thus far of LLMs in linguistics. Then, it proceeds to Piantadosi's critique of Chomsky in particular, as a useful foil for the overall argument. It then identifies CALU as the missing thread in linguistics debates about LLMs. A full explication of CALU is then provided, followed by an illustration of how LLMs fail to replicate it. From there, implications of the foregoing arguments for linguistic theory are teased out, including the necessity of a competence-performance distinction, the internalist orientation of linguistics and the postulation of a modular and generative language faculty, and the excessive ambition of LLMs conceived as theories of human language.

2. Large language models and generative linguistics

Piantadosi's (2023/2024) critique represents the latest episode in a long-running drama about the utility of generative linguistics, particularly as articulated by Chomsky. The piece drew from the momentum generated by the success of OpenAI's (2022) ChatGPT-3.5, to which Chomsky, Roberts and Watumull (2023) had already responded in *The New York Times* in March 2023.

To be sure, the human-like output of LLMs rightfully triggers a rethink of fundamental claims pertaining to themes in generative linguistics. Indeed, Piantadosi is far from alone in this respect—many scholars see direct implications for the study of human language from the rise of contemporary language models (e.g., Beguš et al. 2023; Kallens et al. 2023; Mahowald et al. 2024).

Perspectives vary, with different theoretical weight assigned to LLMs among sympathetic scholars. For example, Futrell and Mahowald (2025) argue that LLMs and theoretical linguistics are mutually beneficial, aiding the development of one another in various respects. In this way, they take a somewhat less strident position than Ambridge and Blything, who argue that LLMs are "*better* theories than any other that have been proposed" (Ambridge and Blything 2024: 34).

More sympathetic to the early theoretical work of Chomsky et al. (2024) argue that levels of adequacy and discovery procedures identified in those works are relevant to the study of LLMs, noting a difference between the statistical learning methods of earlier n-gram (Markov) models and LLM learning today.[2] They later extended their framework in service of the claim that LLMs and generative linguistics are compatible (Portelance and Jasbi 2025).

Others draw a sharper line between the two. Volenec and Reiss (2025, 2-3) argue that proponents of LLM-driven theorizing, like Piantadosi (2023/2024) mistake the research paradigm's orientation toward an account of *linguistic knowledge* (its content and acquisition) for an account of *acceptability judgments*. The competence-performance distinction, they argue, is crucial to generative linguistics yet neglected in critiques, with a distinction between the object of inquiry (the human language faculty) and sources of evidence (acceptability judgments, eye-tracking data, brain imaging, etc.) resulting from it (Volenec and Reiss 2025, 3-5).

2.1 Piantadosi's critique of Chomsky's approach

Piantadosi's critique, given its focus, provides a useful foil for our purposes.

It is worth noting that while principally framed as a direct critique of Chomsky's "approach," Piantadosi's critique frequently veers into the broader territory of generative linguistics. This may be because, as Piantadosi (2024, 385-387) himself acknowledges in the critique's postscript, the nature of *scientific inquiry* seems to be lurking behind arguments between himself and his critics.

The central claims by Piantadosi are (1) LLMs are linguistic theories in their own right, and (2) These theories undermine the Chomskyan approach to linguistics. Piantadosi argues that LLMs, because they "develop representations *of* key structures and dependencies," therefore "should be treated as bona fide

[2] They also emphasize the potential of neural grammar induction models in this context.

linguistic *theories*" (Piantadosi 2024, 360). He is adopting the position of Baroni (2022) here, who argues that it is "appropriate...to look at deep nets as *linguistic theories*, encoding non-trivial structure priors facilitating language acquisition and processing...a general theory defining a space of possible grammars" prior to language-specific training (Baroni 2022, 7).[3]

The claimed refutation of Chomsky's approach can be found in the idea that language models capture human-like representations of syntactic and semantic structure—they reproduce the aspects of natural language that linguists seek to explain (Piantadosi 2024, 362-367). Acknowledging that "any model will necessarily have certain tendencies and biases," Piantadosi thus conceives of "each model or set of modeling assumptions as a possible hypothesis about how the mind may work. Testing how well a model matches humanlike behavior then provides a scientific test of that model's assumptions" (Piantadosi 2024, 364).

In the postscript, Piantadosi (2024, 385) refers to his argument as a form of *inductive* reasoning, *predicting experimental data*, rather than making deductions *from* experimental data. In this way, the argument is that modern language models shed light on human language because they accurately predict the structure of human language via their simulation. This hints at deeper issues of methodology that will be relevant to generative linguistic methodology, and CALU is explicated in this chapter.

2.2 A missing thread

Nevertheless, Piantadosi's critique (and the broader debate) does not touch on the matter of CALU and what that portends for linguistic theory.

Surprisingly, neither do most responses to Piantadosi and other scholars sympathetic to LLM-driven linguistic theorizing. (One notable exception is given by Moro et al., who briefly note that LLMs do not exhibit stimulus-freedom and are thus "obviously lacking creativity in this sense" (Moro et al. 2023, 83).

Perhaps this is not surprising, though. Consider Georges Rey's comment on the matter:

---

[3] Baroni (2022, 6-7) takes a significantly firmer stance than Piantadosi on the matter of innate priors and language models.

> [Chomsky] discusses this issue so frequently, it would be easy to think it's an essential part of this theory, or even what he takes the data for it to be. As we'll see, however, other than the fact that we utter constantly novel utterances, issues of how we use language, and whether it involves "(free) will" and choice, which may or may not be compatible with "contact mechanics," are, in fact, entirely inessential to his core theory and to the crucial data for it (Rey 2020, 39).

Chomsky has, on this interpretation, thrown a wrench into linguistic discourse, simultaneously making the significant claim that the human language faculty is freely exercised *without* making this a central—or even essential—part of his theorizing.

But is this picture accurate?

McGilvray notes to the contrary that Chomsky has long considered CALU to represent a set of facts about ordinary human language use with which "any science of language must contend" (McGilvray 2001, 5). Indeed, Chomsky has sometimes referenced the "Galilean challenge" —the problem of why a "machine may be impelled to act in a certain way, but it cannot be inclined; with human beings, it is often the reverse" (Chomsky 2017, 1).[4] The establishment of general computability theory in the early-to-mid twentieth century meant that it "became possible, for the first time, to address part of the Galilean challenge directly, even though the earlier history remained unknown" (Chomsky 2017, 1-2).

Chomsky is likewise clear that certain theoretical constraints central to generative linguistics resulted from this reformulation of the problem of ordinary language use. This includes distinguishing between "the internalized system of knowledge [of language] from the processes that access it" (Chomsky 2017, 2). That is, between *competence* and *performance.* "The theory of computability enables us to establish the distinction, which is an important one, familiar in other domains" (Chomsky 2017, 2). The orientation is *internalist* (more on this below).

Linguistic performance—taken in its full breadth, comprising CALU—presents a set of facts that must be dealt with by an explanatory framework worth its salt; the "why" question of linguistic theory (why it is one way and not another) cannot be properly framed without taking it into account, in some way.

---

[4] Rey (2020, 387-388) is, for his part, cool to this notion of behavior that is 'inclined but not impelled.'

What *is* inessential for Chomsky is an explanation for CALU itself. Chomsky holds that linguistic performance is likely beyond the scope of current and potentially future scientific explanation. His emphasis on addressing "part" of the Galilean challenge meant a reformulation of a problem traceable back to Descartes (1637) and Cordemoy (1668), who held that ordinary language use distinguishes humans (minded creatures) from machines (lacking minds). The rise of computability theory provided the tools to make the reformulation; to conceive of the infinite generativity of a finite system (the brain) and posit an idealized "neurobiological Turing machine" (Watumull and Chomsky 2020, 4)

Thus, one can, for example, engage with Chomsky's articulations of the Minimalist Program or his thoughts on the evolution of the language faculty without any engagement with this notion of free language use, as none of these more familiar subjects requires direct engagement with the problem of uncaused but appropriate language use.

Yet, this does not indicate that CALU is not part of the foundation of Chomsky's linguistic theorizing. The set of facts bound up in CALU has downstream implications for a science of language.

3. The creative aspect of language use

A fuller explication of CALU is useful to further flesh out its role in a science of language. While some responses to potential objections to the idea are provided, the focus here is primarily on drawing out its implications for linguistic theory, rather than a comprehensive defense of the idea.[5]

Chomsky (e.g., 2009, 69-77) often cites a phrase attributable to Wilhelm von Humboldt: *infinite use of finite means*. This phrase can be parsed in (at least) two complementary ways. The first emphasizes the ability of human beings to construct, with only finite mental resources, an infinite array of structured linguistic expressions. This is the infinite generativity of language use. The second, compatible with the first, emphasizes the *use* of finite means over an indefinite range of both expressions and circumstances.[6] The phrase helps get at this idea of what Chomsky means by 'free' language use.

---

[5] For more thorough defenses of the idea, see, Chomsky (1966, ch.1), McGilvray (2001), Asoulin (2013), and Carchidi (2024, 2025).

[6] Popular (and often linguistic) portrayals of this phrase and its association with Chomsky typically emphasize only the first parsing (see, e.g., Summerfield 2023: 249). However, as Chomsky explicitly notes: "[A] number of professional linguists have repeatedly confused

Ordinary language use has the characteristic of being innovative. That is, no individual with a mature and developmentally normal language capacity is confined to a pre-sorted list of sentences. It is *unbounded.*

Additionally, ordinary language use does not appear confined to a pre-defined set of stimuli for which it is recruitable. That is, individuals do not appear strictly reliant on stimuli that one can select in advance to use their language in this innovative fashion. It is *stimulus-free.*

Finally, ordinary language use does not appear to have a pre-specified list of situations for which its use is appropriate to the situation at hand and coherent to others who hear (or otherwise interpret) the remarks. That is, individuals use their language in an unbounded and stimulus-free fashion, though their linguistic expressions converge with the mental states of others. It is *appropriate to circumstance.*

"In a phrase," Collins writes in summary of the idea, "a speaker makes free, infinite use of finite means. It is as if the speaker has harnessed randomness (unpredictable unboundedness) in a meaningful, appropriate manner" (Collins 2006, 560). Each of the three characteristics of language use described can be explicated independently:

Stimulus-Freedom: Animal communication systems are stimulus-bound in that the expression of mental content appears restricted to circumstances (see Anderson 2008, 796-797). Human language use, in contrast, does not appear to occur in some kind of fixed relationship with identifiable stimuli in the local context. It is *detachable* from local, and even *possible* (i.e., fictional), contexts (see Boeckx 2010, 36).

Research purporting to find human-like linguistic characteristics in non-human primates does not appear to claim as much for the latter. For example, a recent study claiming that bonobos exhibit "nontrivial" compositionality (the ability to combine elements with independent meanings into new combinations in which these elements modify one another) indeed defines the meaning of a bonobo's signal "as the set of features of circumstances (FoCs) that appear at a rate greater than chance across the signal's occurrences" (Berthet et al. 2025,

---

what I refer to here as "the creative aspect of language use" with the recursive property of generative grammars, a very different matter" (Chomsky 2006: xiv). This is not to say that Chomsky's interpretation of Humboldt's work (being an ancestor of modern generative grammar) is correct, but to say that both parsings are used in his work.

104). Such bonobo signals are therefore explicitly tied to the circumstances of their use, at least in this study's framework.

Human language use is not like this. Instead, ordinary language use is *elicited* by local stimuli (we are, say, inclined to respond to remarks directed at us) but not *caused* by them.

Some are chilly to the general idea. Harman (1968, 231-234) argued that both internal and external stimuli *do* control language use. For external stimuli, he argued, language comprehension by a hearer "is, or ought to be, a function of what is said" (Harman 1968, 232). For internal stimuli, the assertion of stimulus-freedom amounts merely to ignorance of the causal factors underlying linguistic performance (Harman 1968, 230-232). On this latter point, Rey (2020) levels a similar objection, noting that "the current natural sciences of human bodies seem to indicate that their states and motions are not only "inclined," but every bit as "compelled" as the states and motions of any machine" (Rey 2020, 387).

On the first point, Chomsky's position on external stimuli traces back to a least his 1959 critique of Skinner's (1957) *Verbal Behavior*. In response to the notion of *stimulus control*, Chomsky says: "We cannot predict verbal behavior in terms of the stimuli in the speaker's environment, since we do not know what the current stimuli are until he responds" (Chomsky 1959, 32). That is, identification of a given stimulus to which language use is allegedly fixed (caused by) depends on the individual using language arbitrarily, at which point—and only that point—can the scientist identify a stimulus in the environment.

When Chomsky accuses the behaviorist conception of stimulus control as being merely a guise for "mentalistic psychology" (Chomsky 1959, 52), he is drawing attention to the intrinsically *post-hoc* effort to explain language use. To try and account for language use by appealing to properties in the physical environment, as Skinner did, "the world *stimulus* has lost all objectivity in this usage. Stimuli are no longer part of the outside physical world; they are driven back into the organism" (Chomsky 1959, 52). Indeed, there appears to be no "fixed association of utterances to external stimuli or physiological states" (Chomsky 2009, 60).

Any attempt to provide a serious causal explanation for an individual's linguistic production will merely be an *interpretation*—and not a scientific one—of many factors about the given context—"[t]his is not the well-defined causality of serious theory..." (McGilvray 2001, 7; see also Asoulin 2013, 230). Individuals may, in the face of a potential stimulus (e.g., being asked a question), say nothing at all.[7]

Internal stimuli seem to pose a different problem. The *brain,* accepted wisdom holds, *causes* language use from within. Although we may not be able to *predict* what an individual may (or may not) say in an arbitrary situation, is this tantamount to suggesting language use is detachable from *any* stimulus, including those internal, as though speaking from a void?[8]

To keep our focus, we note immediately that, *even if* human language use were only free of *external* stimuli, this would still mark a disjuncture from computational models, including LLMs, which output in ways controlled by identifiable external and internal stimuli (see below). But how far can the notion of stimulus-freedom be extended for humans?

A dialogue including, among others, Collins and Rey (see Collins 2006, 473-477) on this very point surfaces at least one counter-objection: the *inverse* of the original problem pointed out in Chomsky's (1959) review of Skinner arises when one tries to locate an internal stimulus. That is, if this assumption is accurate—that language use must have an internal, causal factor owing to some complex interplay of physiological mechanisms—then it is (at least as a reasonable standard of scientific inquiry) incumbent upon the scientist to identify such and such internal event that leads to such and such linguistic expression. Yet, at this point, the scientist is merely pairing up a kind of "putative" cause with a given utterance, rather than offering any systematic explanation (Collins 2006, 474).

The problem here, without straying too far from our purpose here, is that the rush to assume that all human behavior can be understood in 'mechanical' terms tries to do away with the *at will* component of human language use; humans speak in ways they *intend* to speak, expressing thoughts they *intend* to express (in ways conformable to situations). Yet, the assumption does not reach

---

[7] This freedom from stimuli should not be taken to mean that a given stimulus—say, another person's spoken remark—has no controlling effect on how the receiving individual *perceives* that stimulus. To use a familiar example: a native speaker of English cannot help but hear words spoken in English *as English.* They are *not* free of this perception and, indeed, cannot *choose* to be free of it. They *are* free to say whatever they please in response to spoken English (or any other language) or to say nothing at all.

[8] I appreciate an anonymous reviewer's feedback in raising this objection, which is useful for clarity.

the same bar held more clearly in the case of external stimuli: no fixed association of utterances with internal states can be identified.

It also raises the Skinner-esque problem: if my linguistic utterances are generated and spoken through a purely 'mechanical' process, owing to a complex interplay of physiological processes, then why do they converge with your mental state (which is undergoing its own 'mechanical' processes)? In other words, how would stimulus-controlled utterances be meaningfully and appropriately applied to situations?

More than this, attempting to identify a stimulus for an *at will* behavior is nearly a contradiction—the intended expression of thought through language describes a person, not a mechanism. Stimulus-control is simply the wrong concept, inadequate to the task at hand.[9]

Unboundedness: Human language use is novel and innovative owing to its *combinatorics.* This is likely the least controversial characteristic to cognitive scientists. Language use consists of combining and recombining *finite* parts to form *new meanings* without an upper limit on the ability to produce new meanings (Anderson 2008, 797) (save for non-relevant constraints, like memory, fatigue, etc.). The unboundedness of language use thus traces back to language's code—syntax (Moro 2016, 21-23). The ability to produce infinite novel meanings from finite elements in natural language is sometimes glossed as the *infinite productivity* of language (Huybregts 2019, 2).[10]

Appropriateness and Coherence to Circumstances: The stimulus-freedom of human language use is revelatory of the third characteristic: rather than being caused by local stimuli, language use is instead connected to external stimuli "by the much more obscure relation of appropriateness" (Chomsky 1975, 302). Appropriateness is an elusive notion—there is, as McGilvray (2001, 8-18) argues, no theory that is forthcoming that makes it possible to simply list all those conditions in which an utterance is appropriate, as language use is *neither* determined nor random (being stimulus-free and unbounded) *yet* appropriate to others. There is, moreover, no apparent limit on the number of situations in which language use *could* be appropriate.

Individuals nevertheless possess the conceptual resources to judge appropriateness *intuitively* and over an indefinite range of situations. These

[9] For further defense, see Carchidi (2025).

[10] Explanation notwithstanding, the infinite productivity of human cognition is not an observation exclusive to generativists (see, Piantadosi and Fedorenko 2017).

resources are held to be biologically endowed by the generativists. We may make two certain statements, then, about the "appropriateness" conditions. First, appropriateness cannot refer merely to a *functional* use of language in which it is goal-oriented or otherwise context-dependent *because* it is stimulus-free (McGilvray 2001, 8-10). Second, language use is bound up in *expressions of thought.*[11] Chomsky thus refers to individuals as judges of appropriateness: It "is recognized as appropriate by other participants in the discourse situation who might have reacted in similar ways and whose thoughts, evoked by this discourse, correspond to those of the speaker" (Chomsky 1988, 5).

Only *together* do these three components constitute CALU (Baker 2007, 236-237).

4. LLMs do not replicate CALU

The natural next question is: do LLMs reproduce CALU?

Let us consider each characteristic:

Stimulus-Controlled: Unlike the problem pointed out in Chomsky's (1959) review of Skinner, a set of stimuli that control the outputs of LLMs can be identified. LLMs' outputs[12] are stimulus-controlled in two primary ways: (1) Given an input, the LLM will generate an output value, unless instructed to the contrary (which reinforces the point); and (2) The LLM will not decide to *not* generate an output value, to *alter* the process of token generation (putting aside the specifics of the operations it performs to transform an input into an output), nor to *initiate* its own conversations *without* instructions to do so.

LLMs' outputs are inextricably tied to the "environment" established by their internal programming and external prompting. While an LLM could be manipulated by human programmers in such a way as to make its stimulus responses variable (e.g., to not respond to every third input), the LLM is

[11] Responses to Chomsky arguing that language is primarily a tool for communication rather than thought, most prominent among recent examples being Fedorenko et al. (2024), wholly neglect Chomsky's characterization of creative language use and what this might mean for its conceptualization as a "tool" (which would assign a function to something that appears not to have pre-defined uses, being stimulus-free). See Carchidi (2025, Forthcoming) for elaborated responses.

[12] We are speaking about the base model of any conversational system (e.g., ChatGPT), though the point applies equally to applications underpinned by LLMs (see, on this structure, Shanahan 2023, 4).

ultimately bound to either the internal state imposed by human programmers or the prompt given to the model (or a combination of both).

To be clear, to say that an LLM's outputs are stimulus-controlled is *not* to say that the process it employs to transform an input value into an output value is determined, say, by an external human prompter requesting this or that response. That said, the stimulus-controlled nature of its outputs indicates that the LLM's architecture lacks a kind of operational autonomy; its resources are simply there for the taking, so to speak. As we see below, this is quite unlike the cognitive architecture necessary to enable human stimulus-freedom, namely, a modular and generative language faculty.[13]

LLMs exhibit no independent ability to respond or not respond; they are impelled to act but not inclined.

Weakly Unbounded: LLMs exhibit a weak form of unboundedness, relative to human linguistic competence. They can generate new sentences according to whatever model of a given dataset(s) they acquired during training, according to a particular configuration (see generally, McGilvray 1999, 85-86). Syntactic novelty has been found experimentally in GPT-2 (McCoy et al. 2023).

However, the unboundedness of CALU is strong in the sense that the meanings of linguistic expressions and their grammatical structures are understood by humans. Human language affords the "intricate regulation of form/meaning pairs," Murphy and Leivada write, and it is this that "constitutes the stuff of syntactic theory, not the organization of strings" (Murphy and Leivada 2022, 1). Several reasons indicate LLMs do not meet this stronger interpretation, including those detailed in contributions to this volume.

An obvious example is the persistence of "hallucinations" in LLMs—outputting false or wholly fictional outputs authoritatively. The comparison may seem skewed against LLMs, as human beings, after all, both err and lie. But in computational models like LLMs, mistakes cannot be attributed to various

[13] This is a *weaker* form of the argument. Colin McGinn (2021), somewhat informally, has argued that the production of linguistic expressions that convey one's thoughts is *necessarily* stimulus-free and, conversely, any organism *lacking* stimulus-freedom cannot engage in the infinite productivity of human language *because* they are bound to stimuli (the expressions being more akin to the reflex of the knee when tapped). I remain agnostic to this argument here, though the reader should at least bear it in mind: it would entail that stimulus-controlled LLMs *necessarily* lack the ability to meaningfully structure elements into sentences, which would limit their ability to engage in the *unboundedness* of human language.

performance constraints suffered by humans—there are no comparable distractions that may cause their concentration to break, no fatigue from sleep deprivation that may lead to inaccurate statements, and so on.[14] For humans, these contribute to such errors.

LLMs likewise do not lie, as Moro et al. (2023, 83) note. For humans, lying is not an indication that they cannot properly regulate form/meaning pairs, but an indication that they can, in fact, use this regulation towards ends unrelated to language. LLMs have no such recourse.

There are also the *kinds* of hallucinations LLMs exhibit. One April 2025 study, for example, noted the discovery of a "generalization bias" in state-of-the-art LLMs tested in clinical trial results summarization tasks. The conclusions of the trials were prone to over-generalization. Humans, as the authors note, are prone to this as well. Yet, amusingly, when nine of ten models[15] tested were explicitly prompted to keep their responses accurate, their resulting outputs were "about twice as likely" to contain generalized conclusions (Peters and Chin-Yee 2025, 10). Whatever the proper interpretation of this is, it does not appear to resemble human behavior.

Other, perhaps less appreciated reasons, dovetail nicely with this line of thought. As David Lobina (this volume) details in Chapter 3, adversarial attacks on LLMs—perturbing the input data of a model in subtle, often bizarre (to humans) ways—effectively 'disrupts' the LLM's comprehension, whereas humans tend to remain robust to the changes. This does not indicate that LLMs are regulating form/meaning pairs over an indefinite range but instead serves their intended purpose as function approximators (Chapter 3, this volume).

---

[14] Interestingly, testing LLMs on the same problems but on different hardware (GPU clusters) caused result instability in one set of experiments testing models' performance stability on mathematics benchmarks (see Hochlehnert 2025, 7-8). The results seem to cast doubt on the remarks in the main text. Ironically, the best way to interpret them is with a competence-performance distinction, as the same model performing differently on the same problems across different GPU clusters does appear to be a performance constraint, rather than a lack of competence. But the *kind* of inaccuracy exhibited in these results is inhuman, as any human who otherwise grasps the logic of a set of problems tends not to err the same ways (unless distracted, and so forth). There is also no comparison to be made with humans for the swapping out of underlying "hardware" (a metaphor, at best).

[15] A Claude model was the one exception to this result.

Perhaps also less appreciated is that computational models' training procedures are *idealized* relative to human language acquisition, as Dupre (2024) argues. Principally concerned with grammar induction models, Dupre argues that, to be considered adequate models of human learners, models must somehow 'factor out' the myriad causal sources beyond language itself (communicative goals, externalization systems, etc.) that comprise a human learner's dataset.

The point for LLMs is that their usefulness as theories likewise depends on providing an alternative account of multiple sources of data on human language acquisition. LLMs may not clear the bar in this respect, and any resulting competencies attributed to them may mistakenly be conflated with human linguistic competence.

Finally, "reasoning" models—LLMs with significant post-training innovations that lead them to output candidate "chains-of-thought" —throw a curveball. The chains of thought resemble the verbalized thought processes of humans. Indeed, the DeepSeek-R1 research paper reports an "aha moment" during DeepSeek-R1-Zero's training: the researchers described the model as having learned to "allocate more thinking time to a problem by reevaluating its initial approach" (DeepSeek-AI et al. 2025, 8) (the model outputs, when trying to solve an equation, "Wait, wait. Wait. That's an aha moment I can flag here") (DeepSeek-AI et al. 2025, 9).

The output is remarkably human-like. The perception that this fuels the acquisition of a general-purpose reasoning capability underlies much of the enthusiasm for these models' development.

However, such outputs may be misleading. One team of researchers from Arizona State University trained and tested three transformer models on the traces and solutions of an algorithm designed for grid-traversal and path-finding (A* search). Once trained, the models are tasked with constructing plans to navigate a grid and reach an endpoint via legal actions (up, down, left, and right). Bizarrely, the model with the best performance—which showed slightly better out-of-distribution performance, even—was the one trained on A* search chains-of-thought that were deliberately corrupted through random mismatching by the researchers! The implication is that chains of thought are not predictive of final outputs (Stechly et al. 2025).

The results indicate that LLMs have not acquired a linguistic capacity that regulates form/meaning pairs, so much as the *model is simply doing something different and unhuman-like.*[16] (Function approximation seems to be the better lens, here.)

Yet, generative linguists do not take their object of study to be the externalization of speech emulated by chains of thought. Rather, they study an internal competence—an internalized system of knowledge—distinct from the natural language characters outputted in chains-of-thought.[17] That LLM chains-of-thought appear to be causally independent of their final outputs poses only a problem for those who argue that LLMs *are* models of human language and that human language *is* speech or written words. Usage-based approaches specifically have to account for this mismatch. The generative linguist is focused on a different matter, for which LLMs may simply have little to no relation.

LLMs are weakly unbounded, relative to humans.

Functionally Appropriate: LLMs, by virtue of being stimulus-controlled, immediately fail to exhibit the appropriateness criterion. They do exhibit a kind of *functional appropriateness* in the sense that their outputs exist in a functional relationship with their internal programming (any hand-coded instructions provided to it, for example) and external environment (the prompter). Efforts to align the LLM with the expectations of a human end-user indeed result only in the restriction of its outputs to serving goals established in its local environment by the human programmer.[18]

Yet, human language use is *elicited* by local stimuli, not caused by them. LLMs' outputs are not elicited by local stimuli but *caused* by them.

---

[16] Although the A* search traces used to train the transformers are not strictly linguistic, the point is fundamental: the prior chain of thought does not appear causally responsible for the final output.

[17] Of course, as Mendívil-Giró notes in Chapter 1 (this volume), LLMs output tokens, not natural language characters (see also Manova, Chapter 6, this volume).

[18] McGilvray marks a distinction here: creative linguistic productions are stimulus-free, so "being appropriate" is not "being caused by environmental conditions." Nor is it being regularly correlated with the environment" (McGilvray 2001, 9). He explains that this "is not to say that regular correlation, with or without causation, yields no conception of appropriateness at all" (McGilvray 2001, 9). Creative appropriateness is not "functional," goal-oriented appropriateness and one cannot capture this criterion merely through "articulate scheme(s) or formula(e)" (McGilvray 2001, 9). One need not satisfy a standard or speak in reference to an established goal in one's environment for their remarks to be appropriate.

LLMs, in sum, *do not exhibit* the criteria sufficient to designate their outputs as "creative" use, in our sense.

5. The Implications of CALU for LLMs and Linguistic Theory

That LLMs do not replicate CALU indicates that they are not *language users* in the sense described here. They do not *use* language in a creative manner operationalized in terms of stimulus-freedom, unboundedness, and appropriateness.

To be sure, debates about LLMs in linguistics tend to set aside this designation of "language user," mirroring CALU's relative absence in much discussion of generative linguistics. Contreras Kallens et al. (2023, 3), for example, while arguing that LLMs demonstrate the significance of statistical learning, explicitly rule out the possibility that LLMs are language users because they lack the requisite cognitive capacities for social interaction. Mahowald et al. (2024, 536) consider the question of language use to be an outstanding question. Piantadosi (2024, 38) conceives of LLMs as simulations of human language, which appears to rule out their function as language *users.*[19]

The implications of LLMs' lack of creative language use ought to be teased out, beginning with the theoretical constraints necessary to make sense of the cognitive capacity of a human language user.

5.1. Competence and performance

The competence-performance distinction is often raised in the context of human behavior's *variability*; it is not feasible to draw conclusions about the human language faculty based on variable behavior, thus a distinction is drawn between competence (linguistic knowledge) and performance (linguistic behavior). Resulting from this competence-performance distinction is a further distinction between the object of inquiry (the human language faculty) and sources of evidence (acceptability judgments, brain imaging, etc.). Sources of evidence must be *controlled*, and have their variability focused, to properly study the language faculty (see Reiss and Volenec 2025, 4-5).

---

[19] Lederman and Mahowald (2024) do, to be sure, make the case that the novel text-generation of LLMs does pose a challenge to, among other things, the idea that LLMs are merely cultural technologies for the transmission of existing knowledge. However, this operates too far afield from our focus to be relevant here.

To be sure, this is but one route to the competence-performance distinction. Even among some generative linguists, the thinking normally goes that competence is what is being studied, therefore performance—which is not the object of inquiry—is (temporarily at least) irrelevant. However, this neglects that the set of facts bound up in CALU represents another route to the competence-performance distinction. CALU is both a source of said variability and raises, apart from other matters of behavioral variability, deeper issues about linguistic inquiry.

We noted above that Chomsky was influenced by work in the formal sciences in the early days of generative linguistics and the establishment of general computability theory in the early twentieth century. Chomsky and colleagues took stock of Turing's insight that "[i]t is possible to invent a single machine which can be used to compute any computable sequence" (Turing 1937, 241); the infinite generativity from finite means of central importance to generative linguistics.

Indeed, as Mendívil-Giró details, Universal Grammar was formulated to conceive of language as natural, biological object characterized by *recursive enumerability* which is capable of *digital infinity* (Mendívil-Giró 2018, 861). This computational system recursively produces grammatical sentences over an indefinite range (Mendívil-Giró 2018, 873-874). This finite object "characterizes an infinite array of 'free expressions,' each a mental structure with a certain form and meaning" (Chomsky 2007, 45). Universal Grammar is the study of its innate basis (Chomsky 2007, 45-46). To the point, the term *recursion* was directly inspired by work in mathematical logic, which made the term "*recursive*... equivalent to *computable*" (Mendívil-Giró 2018, 874).

Yet, this computational system (the language faculty) did not alleviate the earlier Cartesian problem of creative language use in production. Internalized knowledge capable of yielding digital infinity is a separate matter from the externalization of that knowledge, though the distinction makes sense only if one can justifiably conceive of infinite generativity from a finite system. The postulation of an inborn computational system meant that it "became possible, for the first time, to address part of the Galilean challenge directly, even though the earlier history remained unknown" (Chomsky 2017, 1-2).

Part of the challenge. The *use* of this computational system in production is another matter, for which a distinction must be drawn: the abstract knowledge encoded in this system (competence) and the use of this system in production (performance).[20] The level of the individual is still present, so to speak, not resolved by the tools of computability theory (see generally Carchidi 2025, Forthcoming).

There are more serious issues at play. CALU deals with the uncaused but appropriate expression of thought evidenced through ordinary natural language use. Behavior that is stimulus-free is not *deterministic*, though the utterances—often novel (unboundedness)—are not *random* (appropriate). Being neither determined nor random *yet* appropriate, Asoulin notes that the actual *use* of language represented by CALU is "a scientifically intractable aspect of the world" (Asoulin 2013, 235).

Thus, one requires an explanatory theory that seeks not to explain the *use* of language in production, but the mechanisms that enable the individual to generate an unbounded array of structured expressions. This approach to inquiry is *internalist* (see generally, Asoulin 2013). Such a project is only possible if one can conceive of a finite biological system (the human brain) yielding unbounded structured expressions. Computability theory filled this gap, allowing part of the original problem of creative language use to come under the heel of science, hence Universal Grammar's object of inquiry being the innate *basis* of CALU.

CALU, then, is fundamental to a theory of human language but not itself the target explanation. This tracks with McGilvray's observation that CALU, alongside the poverty of the stimulus argument, constitute "the basic observations that lie behind Chomsky's—and other rationalist-romantics'—research strategy or fundamental methodology for the study of language and mind" (McGivlray 2009, 2). He continues: "[T]his suggests that if you want to construct a science of mind and language, you should avoid trying to construct a science of how people use their minds, and especially their language" (McGilvray 2009, 2).

It is not, contra otherwise sympathetic scholars like Rey (2020), that CALU is *inessential* to a science of language. It is that the very contours of linguistic theory depend on these observations about use. To neglect them in theory-formation moves the scholar further from the human being they purport to study.

---

[20] Internalized knowledge reflects what is called I-language whereas externalized speech is called E-language.

LLMs, as noted, do not reproduce CALU. Insofar as LLMs are simulations of human language (per Piantadosi 2024) or an "existence proof" of statistical language learning (per Contreras Kallens et al. 2023, 2) or theoretically astute devices (per Ambridge and Blything 2024), they do not shed any additional light on the stimulus-free, unbounded, yet appropriate use of human language, including any potential explanation of the full problem. As Piantadosi (2024, 367) correctly notes, they make no competence-performance distinction during training, which can retroactively be applied to the human's LLM-theory-of-choice.

Yet, if a competence-performance distinction is derived on the basis of facts about ordinary human language use, and LLMs do not reproduce or invoke it, then their ability to serve as models of human language is diminished. The competence-performance distinction is too important for any theory to neglect.

This does not rule out the use of LLMs in linguistics, but it tempers their usefulness.

5.2. Stimulus-Freedom and the Modular Language Faculty

Further implications can be drawn, extending the foregoing remarks. One bears on the nature of the language faculty under investigation based on the internalist disposition. A central postulation of early generative linguistics was grounding the idea that humans possess an inborn "faculty" of the human brain/mind (whether localized in the brain was not a determinant) that yields infinite generativity with only finite elements. This postulation was made in part on the basis of applying new intellectual tools to an old problem. Is this postulation justified?

Like the competence-performance distinction, arguing about whether the cognitive architecture underpinning human language is domain-specific or domain-general, and related matters, can proceed in CALU's absence. However, the facts bound up in CALU do have downstream implications here, too, particularly from the *stimulus-freedom* of language use.

Stimulus-freedom in language use indicates that the regulation of form/meaning pairs is being carried out by the mind/brain without reliance on identifiable stimuli; there is no pre-defined set of causes in response to which the language faculty is mechanically set into motion. An individual may simply use language. This is the problem of 'free access' (Collins 2004, 519-520), the *at will* recruitment and deployment of linguistic resources. (Per the remarks above, the solution to this problem for generative linguists is only partial, characterizing the *enabling mechanism* for this 'free access.')

The cognitive mechanism(s) enabling stimulus-free language use must therefore fulfill two responsibilities: (1) the isolated production of structured linguistic expressions with a discrete set of cognitive resources; and (2) this production must, in some way, interface with other, relevant cognitive systems. A *modular language faculty* capable of unbounded generativity fulfills both. As McGilvray explains:

> The possibility of stimulus-freedom in language use can be seen to result from a modular language faculty. Not just any kind of modularity will do: we need a faculty that utilizes its own resources and with internal prompting produces (in apparent isolation) through its own algorithms items in the form of linguistic expressions that are unique to it and yet that "interface" with relevant other internal biological systems (McGilvray 2001, 12).

Linguistic expressions, often novel in nature, are generated without fixed reliance on external stimuli. A faculty that specializes in the production of structured linguistic expressions must therefore be present to generate such expressions should its resources be recruited. This system, though modular in the sense that it is independently constituted, must also interface with other cognitive systems relevant to the expressions under production (or else the appropriateness of the remark under production would be lacking).

A simplistic input-output system that takes an input value, performs an operation over the input, and deterministically transforms it into an output value would be insufficient to account for this kind of operational autonomy in language use. This is unsurprising given that LLMs, failing to reproduce stimulus-freedom, operate in precisely this manner (the lack of modularity in the LLM architecture as a problem for linguistics has been noted in a more traditional context by Fox and Katzir 2024).

One could argue that domain-general mechanisms are suited to enabling stimulus-freedom. However, this idea (as though linguistic resources were bounced around the mind mechanically) cannot resolve the matter, as it not only fails to acknowledge the 'free access' of linguistic resources in production but also the isolated nature of linguistic resources this entails. Unless one can find a fixed set of associations for language use, either internal or external, then they cannot be conceived as part of the 'push-and-pull' between cognitive systems. More than this, once broadened to the full scope of ordinary language use, such use must likewise account for the unboundedness and appropriateness of stimulus-free utterances.

This is where the need for this modular language faculty to interface with other cognitive systems arises. The faculty must possess "conceptual resources" that are made "legible" to other systems (McGilvray 2005, 218). Conceiving of the language faculty as a sub-system of the mind/brain is par for the biolinguistic course (see, e.g., Chomsky 2013, 35). It is notably a break from the Cartesian approach to the study of mind from which CALU is derived, in which the productivity of human cognition is owed to a single source (Asoulin 2016: 8).

5.3. LLMs-as-Theories Are Too Ambitious

Chomsky's inclusion of CALU in scientific theory places limits on the scope of its inquiry. Specifically, the target of explanation ought to be the internal cognitive mechanism (competence) that enables linguistic behavior (performance). Behavior itself—in its entirety—is taken to be beyond the scope of the tools that allow linguists to study a finite computational system. It likewise appears to be *neither* determined *nor* random *yet* appropriate, placing it outside the bounds of current scientific inquiry wholesale.[21] Without such limits, a theory of human language risks being overambitious, perhaps driven by a deficient understanding of the problems at hand.

The seeming rigidity of generative linguistic theory, at least as espoused by Chomsky, is sometimes taken to be part of a pattern in which, when confronted with disconfirming data that threatens his theories, Chomsky clings to the core tenets of Universal Grammar. His invocation of the "Galilean" method (or "style") is often interpreted as such (see, e.g., Behme 2014).[22]

Interestingly, Piantadosi (2024, 379-380) takes LLMs to be a refutation of the Galilean method.[23] His extended remarks are provided for clarity:

---

[21] Whether this is a problem in principle or merely owing to current limitations, we put aside here.

[22] Behme incorrectly attributes the term "Galilean" style to Chomsky, claiming he "coined a term for this unorthodox methodology" (Behme 2014: 687). The term originates in Husserl (1970/1936: 41).

[23] For this claim, Piantadosi's (2024, 379) only engagement with Chomsky is a 2012 interview between Chomsky and Yarden Katz (see, Katz, 2012). No scholarly works of Chomsky's are referenced to support this claim about the Galilean method.

> Often, the only way to study such complex systems is through simulation. We often can't intuit the outcome of an underlying set of rules, but computational tools allow us to simulate and just see what happens. Critically, simulations *test the underlying assumptions and principles* in the model: if we simulate traders and don't see high-level statistics of the stock market, we are sure to have missed some key principles…We don't get a direct test of principles because the systems are too complex. We only get to principles by seeing if the simulations recapitulate the same high-level properties of the system we're interested in. And in fact the *surprisingness* of large language models' behavior illustrates how we don't have good intuitions about language learning systems (Piantadosi 2024, 380, emphases in original).

Systems like human language are so complex that the optimal way to test the principles embedded in them is by simulating their use. LLMs perform this simulation, building an internal representation of textual (and other) data based on next-token prediction that gives rise to emergent phenomena. Their success in reproducing human-like language is powerful evidence against the idea that innate, high-level linguistic principles are necessary to give rise to the properties of language in which linguists are interested. From this, Piantadosi believes the Galilean method is refuted.

Piantadosi's remark bears on the relationship between data and theory, decidedly bent towards broad data coverage as a means of explanatory depth. But no real explication is provided of the "method" purportedly employed by Chomsky.

Chomsky's explication of the Galilean method bears on an analogy he makes with CALU. In his 1980 *Rules and Representations*, Chomsky (1980, 8) cites physicist Steven Weinberg's (1976) "The Forces of Nature," which describes the Galilean method in physics as the process of distancing oneself from commonsense ideas about natural phenomena (*abstracting* away from them) and *lowering one's expectations* about what science can realistically accomplish.[24]

---

[24] The emphasis on abstraction and the role of data in the Galilean method is often misunderstood by empiricist-leaning critics. Chater et al. (2015), for example, argue that Chomsky misses two critical principles of Galileo's theorizing: "first, that we must look not to books but to Nature, the real phenomena, if we are to understand the world, and second, the language in which Nature is written is mathematical, which is to say, quantitative in character" (Chater et al., 2015, 96). This emphasis on "real phenomena" and their

Chomsky argues for an analogy between this Galilean style in physics and the method employed by Cartesians: that free human action cannot be meaningfully studied but its enabling mechanisms and influences *are* amenable to scientific inquiry (Chomsky 1980: 8). He suggests that the successes in physics may be "a remarkable historical accident resulting from chance convergence of biological properties of the human mind with some aspect of the real world" (Chomsky 1980, 9).

The driving idea behind this analogy is that CALU may *not* have a "chance convergence" of this kind; our science appears inadequate to deal with it. Thus, the intractability of CALU places limits on what a science of language can realistically accomplish. The Galilean method reinforces, by way of analogy, the distinction between competence (enabling mechanisms) and performance (free linguistic behavior).[25]

All this leads to the conclusion that LLMs, by virtue of not postulating any of these theoretical constraints, nor identifying possible theoretical limitations necessitated by CALU, are *too ambitious.* Adopting them as theories, or fundamentally informative for explanatory theory, risks the illusion of progress.

Adopting LLMs as theories of human language fails to grapple with the *kind* of problem that ordinary language use poses and what this entails for a science of human language. A failure to posit a competence-performance distinction, a resistance to internalism and domain-specificity for language's cognitive basis, and so forth, all flow downstream from a theory that puts excessive stock in LLMs. Mahowald et al., for example, note in a sidebar that whether LLMs are "individual language users or…distributions over potential user outputs" is an open question (Mahowald et al. 2024, 536). The uncertainty here is reasonable. However, there is no engagement with the problem of uncaused but appropriate linguistic behavior; this effort to conceptualize LLMs in comparison to humans misses a fundamental piece of the puzzle.

The general push for LLMs-as-theories also risks a retreat into *commonsense* (or what philosophers call *rational*) explanations for linguistic behavior. Excessive reliance on LLMs for explanatory theory-formation may lead one to assume humans use their language in obvious ways, not requiring fundamental

---

mathematical nature mistakes the Galilean method's disposition as one towards data when it is a disposition towards *abstraction* (see, Collins 2023, 5; Rey 2020, 20).

[25] For more extended remarks on this, see, Carchidi (2024, 18-20). For a comprehensive explication of the Galilean style in generative linguistics, see, Collins (2023).

constraints on one's inquiry. In our personal lives, we ordinarily appeal to these commonsense explanations for others' behavior ("Jim only said that because he's still mad about what Lucy said the other day"), yet this does not clear the bar for scientific inquiry. As McGilvray (2001, 13-18) aptly explains, this would be like dropping the pretension of "H_2O" and using, instead, our commonsense notion of "water" in chemistry.

The virtue of the Galilean method, and the analogy Chomsky makes between the Cartesian thesis on free human action and modern physics, is its move *away* from commonsense ideas about human language. It is not obvious what we are dealing with here. It is likewise not obvious how LLMs offer a depth of understanding previously unattainable. The appropriateness of stimulus-free and unbounded human language use—if it is to be explained—cannot have the "rational," everyday account that we provide for it, just as chemists think of water not as the liquid one drinks but as "H_2O." Nor can it be ignored.

None of this rules out the use of LLMs in linguistics as useful tools. Nor should this be viewed as an empty gesture to proponents of their use as tools of science. Their use, however, must be informed by a theory of human language constructed independently. An immediate implication of this is that, based on the theory one adopts, they will be more useful in some places and less useful in others; they are not theoretical juggernauts, capable of resolving every outstanding problem in linguistics or the broader cognitive sciences. Where LLMs do inform theories, it will likely be second-order in nature; not challenging, but instead *adhering to* the fundamental constraints necessitated by the most basic facts about human language.

6. Conclusion

Perhaps the greatest misconception about Chomsky is that his linguistic work is emblematic of a new scientific method, abruptly departing from the scientific status quo in favor of his own idiosyncratic and ahistorical conception of Universal Grammar. As we have seen, CALU—an old idea—plays a central role in generative linguistic theory, both setting limits on explanatory theory and pointing scholars in directions resulting from those limits. It provides the shape of linguistic theory.

LLMs are likely to continue generating interest in linguistics and the broader cognitive sciences. However, they do not reproduce nor provide insight into CALU previously unavailable to scholars. At best, they are tools to enhance the productivity of linguistic inquiry grounded in theory. At worst, their uses convince scholars that hard problems, like ordinary language use, are no bother at all. This chapter has made the case that their uses are limited and any theories built with or without them require those stubborn, old ideas deeply rooted in generative linguistics.

Bibliography

Ambridge, Ben, and Luke Blything. 2024. “Large Language Models Are Better Than Theoretical Linguists at Theoretical Linguistics.” *Theoretical Linguistics* 50 (1–2): 33–48. https://doi.org/10.1515/tl-2024-2002

Anderson, Stephen. 2008. “The Logical Structure of Linguistic Theory.” *Language* 84 (4): 795–814. https://doi.org/10.1353/lan.0.0075

Asoulin, Eran. 2013. “The Creative Aspect of Language Use and the Implications for Linguistic Science.” *Biolinguistics* 7: 228–48. https://doi.org/10.5964/bioling.8963

Asoulin, Eran. 2016. “Language as an Instrument of Thought.” *Glossa: A Journal of General Linguistics* 1 (1): 1–23. https://doi.org/10.5334/gjgl.34

Baker, Mark C. 2007. “The Creative Aspect of Language Use and Nonbiological Nativism.” In *The Innate Mind, Volume 3: Foundations and the Future*, edited by Peter Carruthers, Stephen Laurence, and Stephen Stich, 233–53. Oxford: Oxford University Press. https://doi.org/10.1093/acprof:oso/9780195332834.003.0012

Baroni, Marco. 2022. “On the Proper Role of Linguistically Oriented Deep Net Analysis in Linguistic Theorizing.” In *Algebraic Structures in Natural Language*, edited by Shalom Lappin and Jean-Philippe Bernardy, 1–16. Boca Raton, FL: CRC Press. https://doi.org/10.1201/9781003205388-1

Beguš, Gašper, Mateusz Dąbkowski, and Russell Rhodes. 2023. “Large Linguistic Models: Analyzing Theoretical Linguistic Abilities of LLMs.” *arXiv*, 1–28. https://arxiv.org/abs/2305.00948

Behme, Christina. 2014. “A ‘Galilean’ Science of Language.” *Journal of Linguistics* 50 (3): 671–704. https://doi.org/10.1017/S0022226714000061

Berthet, Maël, Martin Surbeck, and Simon W. Townsend. 2025. “Extensive Compositionality in the Vocal System of Bonobos.” *Science* 388 (6742): 104–8. https://doi.org/10.1126/science.adv1170

Boeckx, Cedric. 2010. *Language in Cognition: Uncovering Mental Structures and the Rules Behind Them.* Malden, MA: Wiley-Blackwell. https://doi.org/10.1002/9781444310047

Carchidi, Vincent J. 2024. “Creative Minds Like Ours? Large Language Models and the Creative Aspect of Language Use.” *Biolinguistics* 18: 1–31. https://doi.org/10.5964/bioling.13507

Carchidi, Vincent J. 2025. “Rescuing Mind from the Machines.” *Philosophy Now* 168: 1–7. https://philosophynow.org/issues/168/Rescuing_Mind_from_the_Machines.

Carchidi, Vincent J. Forthcoming. "Computational Brain, Creative Mind: Intellectual Freedom as the Upper Limit on Artificial Intelligence." *OSF Preprints*, 1–29. https://osf.io/preprints/osf/g7w6u_v1

Chater, Nick, Andy Clark, John Goldsmith, and Amy Perfors. 2015. *Empiricism and Language Learnability*. Oxford: Oxford University Press. https://doi.org/10.1093/acprof:oso/9780198734260.001.0001

Chomsky, Noam. 1959. "A Review of B. F. Skinner's *Verbal Behavior*." *Language* 35 (1): 26–58. https://doi.org/10.4159/harvard.9780674594623.c6

Chomsky, Noam. 1975. "Knowledge of Language." In *Language, Mind, and Knowledge*, edited by Keith Gunderson, 299–320. Minneapolis: University of Minnesota Press.

Chomsky, Noam. 1980. *Rules and Representations*. New York: Columbia University Press. https://doi.org/10.1017/S0140525X00001515

Chomsky, Noam. 1988. *Language and Problems of Knowledge: The Managua Lectures*. Cambridge, MA: MIT Press.

Chomsky, Noam. 2006. *Language and Mind*. 3rd ed. Cambridge: Cambridge University Press.

Chomsky, Noam. 2007. "Language and Thought: Descartes and Some Reflections on Venerable Themes." In *The Prehistory of Cognitive Science*, edited by Andrew Brook, 38–66. New York: Palgrave Macmillan.

Chomsky, Noam. 2009. *Cartesian Linguistics: A Chapter in the History of Rationalist Thought*. 3rd ed. Cambridge: Cambridge University Press. https://doi.org/10.1017/CBO9780511803116

Chomsky, Noam. 2013. "Problems of Projection." *Lingua* 130: 33–49. https://doi.org/10.1016/j.lingua.2012.12.003

Chomsky, Noam. 2022. "Genuine Explanation and the Strong Minimalist Thesis." *Cognitive Semantics* 8: 347–65. https://doi.org/10.1163/23526416-bja10040

Chomsky, Noam, Ian Roberts, and Jeffrey Watumull. 2023. "Noam Chomsky: The False Promise of ChatGPT." *The New York Times*, March 8, 2023. https://www.nytimes.com/2023/03/08/opinion/noam-chomsky-chatgpt-ai.html

Collins, John. 2006. "Between a Rock and a Hard Place." *Croatian Journal of Philosophy* 6 (3): 469–503. https://doi.org/10.5840/croatjphil2006636

Collins, John. 2021. "Chomsky's Problem/Mystery Distinction." In *A Companion to Chomsky*, edited by Nicholas Allott, Terje Lohndal, and Georges Rey, 557–66. Hoboken, NJ: John Wiley & Sons. https://doi.org/10.1002/9781119598732.ch36

Collins, John. 2023. "Generative Linguistics: 'Galilean Style.'" *Language Sciences* 100: 101585. https://doi.org/10.1016/j.langsci.2023.101585

Contreras Kallens, Pablo, Rachel D. Kristensen-McLachlan, and Morten H. Christiansen. 2023. "Large Language Models Demonstrate the Potential of Statistical Learning in Language." *Cognitive Science: A Multidisciplinary Journal* 47 (3): 1–6. https://doi.org/10.1111/cogs.13256

Cordemoy, Géraud de. 1668. *A Philosophicall Discourse Concerning Speech, Conformable to the Cartesian Principles*. London: The British Library.

DeepSeek-AI et al. 2025. "DeepSeek-R1: Incentivizing Reasoning Capability in LLMs via Reinforcement Learning." *arXiv*, 1–22. https://doi.org/10.48550/arXiv.2501.12948

Descartes, René. 1910 [1637]. *Discourse on the Method of Rightly Conducting the Reason, and Seeking Truth in the Sciences.* Chicago: Open Court Publishing Company.

Dupre, Geoffrey. 2024. "Acquiring a Language vs. Inducing a Grammar." *Cognition* 247 (105771): 1–12. https://doi.org/10.1016/j.cognition.2024.105771

Fedorenko, Evelina, Steven T. Piantadosi, and Edward A. F. Gibson. 2024. "Language Is Primarily a Tool for Communication Rather Than Thought." *Nature* 630 (8017): 575–86. https://doi.org/10.1038/s41586-024-07522-w

Fox, Danny, and Roni Katzir. 2024. "Large Language Models and Theoretical Linguistics." *Theoretical Linguistics* 50 (1–2): 71–76. https://doi.org/10.1515/tl-2024-2005

Futrell, Richard, and Kyle Mahowald. 2025. "How Linguistics Learned to Stop Worrying and Love the Language Models." *arXiv*, 1–55. https://arxiv.org/abs/2501.17047v2

Harman, Gilbert. 1968. "Book Review: *Cartesian Linguistics.*" *The Philosophical Review* 77 (2): 229–35. https://doi.org/10.2307/2183323

Hochlenhnert, Anna, Harshita Bhatnagar, Vasu Udandarao, Samuel Albanie, Aishwarya Prabhu, and Matthias Bethge. 2025. "A Sober Look at Progress in Language Model Reasoning: Pitfalls and Paths to Reproducibility." *arXiv*, 1–27. https://doi.org/10.48550/arXiv.2504.07086

Husserl, Edmund. 1970 [1936]. *The Crisis of European Sciences and Transcendental Phenomenology: An Introduction to Phenomenological Philosophy.* Translated by David Carr. Evanston, IL: Northwestern University Press.

Huybregts, M. A. C. 2019. "Infinite Generation of Language Unreachable from a Stepwise Approach." *Frontiers in Psychology* 10 (425): 1–9. https://doi.org/10.3389/fpsyg.2019.00425

Katz, Yarden. 2012. "Noam Chomsky on Where Artificial Intelligence Went Wrong." *The Atlantic,* November 1, 2012. Accessed June 28, 2025. https://www.theatlantic.com/technology/archive/2012/11/noam-chomsky-on-where-artificial-intelligence-went-wrong/261637/

Katzir, Roni. 2023. "Why Large Language Models Are Poor Theories of Human Linguistic Cognition: A Reply to Piantadosi." *Biolinguistics* 17: 1–12. https://doi.org/10.5964/bioling.13153

Lederman, Harvey, and Kyle Mahowald. 2024. "Are Language Models More Like Libraries or Like Librarians? Bibliotechnism, the Novel Reference Problem, and the Attitudes of LLMs." *arXiv*, 1–15. https://arxiv.org/abs/2401.04854v3

Mahowald, Kyle, Anna A. Ivanova, Idan A. Blank, Nancy Kanwisher, Joshua B. Tenenbaum, and Evelina Fedorenko. 2024. "Dissociating Language and Thought in Large Language Models." *Trends in Cognitive Sciences*: 1–24. https://doi.org/10.1016/j.tics.2024.01.011

McGilvray, James. 1999. *Chomsky*. Cambridge: Polity Press.

McGilvray, James. 2001. "Chomsky on the Creative Aspect of Language Use and Its Implications for Lexical Semantic Studies." In *The Language of Word Meaning*, edited by F. Busa and P. Bouillon, 5–27. Cambridge: Cambridge University Press. https://doi.org/10.1017/CBO9780511896316.004

McGilvray, James. 2005. "Meaning and Creativity." In *The Cambridge Companion to Chomsky*, edited by James McGilvray, 204–22. Cambridge: Cambridge University Press. https://doi.org/10.1017/CCOL0521780136.011

McCoy, R. Thomas, Paul Smolensky, Tal Linzen, Jianfeng Gao, and Asli Celikyilmaz. 2023. "How Much Do Language Models Copy from Their Training Data? Evaluating Linguistic Novelty in Text Generation Using RAVEN." *Transactions of the Association for Computational Linguistics* 11: 652–70. https://doi.org/10.1162/tacl_a_00567

Mendívil-Giró, José-Luis. 2018. "Is Universal Grammar Ready for Retirement?" *Journal of Linguistics* 54 (4): 859–88. https://doi.org/10.1017/S00222267180 00166

Moro, Andrea. 2016. *I Speak, Therefore I Am*. New York: Columbia University Press. https://doi.org/10.7312/moro17740

Moro, Andrea, Matteo Greco, and Stefano F. Cappa. 2023. "Large Languages, Impossible Languages, and Human Brains." *Cortex* 167: 82–85. https://doi.org/10.1016/j.cortex.2023.07.003.

Murphy, Elliot, and Evelina Leivada. 2022. "A Model for Learning Strings Is Not a Model of Language." *Proceedings of the National Academy of Sciences* 199 (23): e2201651119. https://doi.org/10.1073/pnas.2201651119

OpenAI. 2022. "OpenAI. November 30, 2022." Accessed December 27, 2023. https://openai.com/blog/chatgpt

Peters, Ulrike, and Benjamin Chin-Yee. 2025. "Generalization Bias in Large Language Model Summarization of Scientific Research." *Royal Society Open Science* 12 (4): 1–18. https://doi.org/10.1098/rsos.241776

Piantadosi, Steven. 2023. "Modern Language Models Refute Chomsky's Approach to Language." Preprint. https://lingbuzz.net/lingbuzz/007180

Piantadosi, Steven. 2024. "Modern Language Models Refute Chomsky's Approach to Language." In *From Fieldwork to Linguistic Theory: A Tribute to Dan Everett*, edited by Edward Gibson and Mary Poliak, 353–414. Berlin: Language Science Press. https://doi.org/10.5281/zenodo.12665933

Piantadosi, Steven, and Evelina Fedorenko. 2017. "Infinitely Productive Language Can Arise from Chance under Communicative Pressure." *Journal of Language Evolution* 2 (2): 141–47. https://doi.org/10.1093/jole/lzw013

Portelance, Eva, and Masoud Jasbi. 2024. "On the Goals of Linguistic Theory." *arXiv*, 1–19. https://arxiv.org/abs/2411.10533v1

Portelance, Eva, and Masoud Jasbi. 2024. "On the Compatibility of Generative AI and Generative Linguistics." *arXiv*, 1–21. https://arxiv.org/abs/2411.10533 v2

Rey, Georges. 2020. *Representation of Language: Philosophical Issues in a Chomskyan Linguistics.* Oxford: Oxford University Press. https://doi.org/10.1093/oso/9780198855637.001.0001

Shanahan, Murray. 2022. "Talking about Large Language Models." *arXiv.* https://arxiv.org/abs/2212.03551v5

Skinner, B. F. 1957. *Verbal Behavior.* Cambridge, MA: B. F. Skinner Foundation Reprint Series. https://doi.org/10.1037/11256-000

Stechly, Katarzyna, Karthik Valmeekam, Aditya Gundawar, Vedant Palod, and Subbarao Kambhampati. 2025. "Beyond Semantics: The Unreasonable Effectiveness of Reasonless Intermediate Tokens." *arXiv*, 1–14. https://arxiv.org/abs/2505.13775v2

Summerfield, Christopher. 2023. *Natural General Intelligence.* Oxford: Oxford University Press.

Turing, Alan M. 1937. "On Computable Numbers, with an Application to the Entscheidungsproblem." *Proceedings of the London Mathematical Society*, ser. 2, 42 (1): 230–65. https://doi.org/10.1112/plms/s2-42.1.230

Volenec, Veno, and Charles Reiss. 2025. "Adopting Large Language Models as a Theory of Language Does Refute Chomsky (but Not like You Think)." *SKASE Journal of Theoretical Linguistics*: 1–10. https://www.croris.hr/crosbi/publikacija/prilog-casopis/877599

Watumull, Jeffrey, and Noam Chomsky. 2020. "Rethinking Universality." In *Syntactic Architecture and Its Consequences II*, edited by András Bárány, Theresa Biberauer, Jamie Douglas, and Sten Vikner, 3–24. Berlin: Language Science Press.

Weinberg, Steven. 1976. "The Forces of Nature." *Bulletin of the American Academy of Arts and Sciences* 29 (4): 13–29. https://doi.org/10.2307/3823787

About the contributors

Vincent J. **Carchidi** is a researcher in cognitive science and the philosophy of mind, with an interest in artificial intelligence. His work is interdisciplinary, most recently focusing on foundational issues in the generative linguistics tradition, including voluntary linguistic behavior. He likewise maintains an interest in moral cognition and human rights attitudes, focusing on leveraging the research strategies employed by generative linguistics for moral psychology. Carchidi's work appears in academic outlets including the *Human Rights Review*, *AI & Society*, *Biolinguistics*, and the *Cambridge University Press*. He also has an extensive publishing record on technology policy in outlets including *The Hill*, *National Interest*, *Military Strategy Magazine*, and *Defense One*, as well as with think tanks including the Middle East Institute, the Foreign Policy Research Institute, and the New Lines Institute.

Vittoria **Dentella** is a postdoctoral researcher in the Department of Brain and Behavioral Sciences at the University of Pavia. She holds a PhD in Cognitive Science and Language from Rovira i Virgili University in Tarragona, Spain, which was funded by a European MSCA grant. Her research focuses on the relationship between psycholinguistic theories of cognition and computation, with a focus on the role of computational models in advancing our understanding of human grammars.

Annika **Heuser** is a Ph.D. candidate in linguistics at the University of Pennsylvania, advised by Professor Charles Yang. She has a B.S. and M.Eng. from MIT in computer science and brain and cognitive sciences. She takes a computational approach, often using tools from natural language processing, in researching child language acquisition. More specifically, her research focuses on phonological development and how children learn discrete representations that differ from the continuous surface forms that they hear.

Jordan **Kodner** is a faculty member in the Department of Linguistics at Stony Brook University with additional appointments in Computer Science, the AI Innovation Institute, and the Institute for Advanced Computational Science. His work and teaching center on child language acquisition, language change, and the evaluation of natural language processing models from the perspective of linguistics and cognitive science.

Evelina **Leivada** is a psycholinguist, currently an ICREA Research Professor at Universitat Autònoma de Barcelona. Her research brings together both the cognitive and the sociolinguistic aspects of the ability to acquire and use language through investigating the effects of developmental trajectory on language and cognition. More recently, she has focused on Large Language Models and how they deviate or match human baselines. Supporting an interdisciplinary approach to language, Evelina is Associate Editor for Psycholinguistics in the Diamond Open Access journal *Biolinguistics.*

David J. **Lobina** is the author of *Recursion* (2017, Oxford University Press) as well as of an increasing number of articles on what may be termed the theoretical repercussions of generative linguistics, not least as a Monday Columnist at *3 Quarks Daily.* A former *Marie Curie* fellow at the University of Oxford and a former *Juan de la Cierva* fellow at the University of Barcelona, he has also worked at the universities of Milano-Bicocca and Rovira i Virgili. As a result, he specialises in theoretical linguistics, psycholinguistics, cognitive psychology, and in the philosophies of cognitive science and psychology. He is currently spending much time thinking about the present status of the language-of-thought hypothesis and is quite tempted to rewrite *The Language of Thought* all over again.

Miguel **López-Otal** is a Ph.D. candidate in Computer Science at the University of Zaragoza. He holds a degree in Spanish Philology and an M.Eng. in Computational Linguistics from the University of the Basque Country. He specializes in Transformer-based models and the analysis of their linguistic capabilities. His current research focuses on ways large language models could better support minority languages with limited text corpora.

Stela **Manova** holds a PhD in General Linguistics from the University of Vienna, where she served as a senior researcher and lecturer. Her academic background spans mathematics, computer science, and languages, and she has published extensively across theoretical, computational, and experimental linguistics. Her key works include *Understanding Morphological Rules* (Springer, 2011),

Affix Ordering Across Languages and Frameworks (Oxford University Press, 2015), and *Diminutives Across Languages, Theoretical Frameworks and Linguistic Domains* (De Gruyter, 2024). She has edited and co-edited multiple special issues for the journals *Morphology* (Springer) and *Word Structure* (Edinburgh University Press), and contributed to leading handbooks and encyclopedias, including the forthcoming *Oxford Handbook of Iconicity in Language* (2026). Currently based in Vienna, Dr. Manova serves as CEO of MANOVA AI and Principal Investigator of *Gauss: AI Global*, where she applies her linguistic expertise to artificial intelligence research and development.

José-Luis **Mendívil-Giró** is a full professor of General Linguistics at the University of Zaragoza. His research has focused on the theory of grammar, the philosophy of linguistics, and the nature and extent of language change and diversity. He is the author of five books, including *On Biology, History and Culture in Human Language. A critical overview* (Equinox, 2014, with J.C. Moreno). His research articles have been published in journals such as *Folia Linguistica, Journal of Linguistics, Linguistics, Frontiers in Communication, Glossa, Biolinguistics,* and *Theoretical Linguistics.*

Dan **Milway** is a sessional lecturer of linguistics at the University of Toronto. His research focuses on syntactic theory and has published on the theory of grammatical agreement and on the nature of syntactic modification. He writes irregularly about linguistics, philosophy of science, and politics at his website milway.ca.

Elliot **Murphy** is a neuroscientist, linguist, philosopher and political economist at the Vivian L. Smith Department of Neurosurgery, McGovern Medical School, in Houston, Texas. His research concerns the neurobiological basis of syntax and semantics, using intracranial recordings from epilepsy and brain tumor patients. He is the author of numerous books, including *The Oscillatory Nature of Language*, and his research has appeared in *Nature Communications, PNAS, Journal of Neuroscience*, and *Progress in Neurobiology.*

Lucía **Pitarch** is a philologist (Spanish) and Ph.D. candidate in Computational Linguistics at the University of Zaragoza. She has collaborated in European projects such as *Nexus Linguarum*, focused on enhancing open and transparent multilingual resources, and *4DPicture*, a project paving the way towards improving communication in the oncological environment. Her focus is on computational semantics, with contributions on lexico-semantic relation extraction and computational metaphor processing at top conferences like ACL, ESWC or LREC.

Héctor Javier **Vázquez Martínez** is a Ph.D. researcher in Computational Linguistics at the University of Pennsylvania, with a B.S. and M.Eng. in Electrical Engineering and Computer Science from MIT. His work sits at the intersection of deep learning, natural language processing, and linguistics, specializing in modeling human speech and exploring how it can be represented and acquired by both minds and machines.

Charles **Yang** teaches linguistics, computer science and psychology and directs the Cognitive Science Program at the University of Pennsylvania. He is the author of several books on language learning, variation, and change, including *Knowledge and Learning in Natural Language* (Oxford University Press, 2002), *The Infinite Gift* (Scribner, 2006) and *The Price of Linguistic Productivity* (MIT Press, 2016), which won the *Leonard Bloomfield Award* from the Linguistic Society of America.

Index

E

F

G

H

Q

R

S

T

U

V

W

X

Y

Z

www.ingramcontent.com/pod-product-compliance
Lightning Source LLC
LaVergne TN
LVHW052311200926
848481LV00002B/528

* 9 7 9 8 2 6 1 6 0 0 4 6 6 *